Advances in Environment, Behavior, and Design
VOLUME 3

A Continuation Order Plan is available for this series. A continuation order will bring delivery of each new volume immediately upon publication. Volumes are billed only upon actual shipment. For further information please contact the publisher.

Advances in Environment, Behavior, and Design
VOLUME 3

Edited by

ERVIN H. ZUBE
School of Renewable Natural Resources
University of Arizona
Tucson, Arizona

and

GARY T. MOORE
School of Architecture and Urban Planning
University of Wisconsin–Milwaukee
Milwaukee, Wisconsin

Published in cooperation with the
Environmental Design Research Association

PLENUM PRESS • NEW YORK AND LONDON

Library of Congress Catalog Card Number 88-649861

ISBN 0-306-43643-4

Contributors

Franklin Becker, Department of Design and Environmental Analysis, College of Human Ecology, Cornell University, Ithaca, New York 14853

Louise Chawla, Whitney Young College, Kentucky State University, Frankfort, Kentucky 40601

Carole Després, School of Architecture, Laval University, Quebec G1K 7P4, Canada

Jay Farbstein, Jay Farbstein & Associates, Inc., 1411 Marsh Street, Suite 204, San Luis Obispo, California 93401

Linda N. Groat, College of Architecture and Urban Planning, University of Michigan, Ann Arbor, Michigan 48109

Thomas C. Hubka, Department of Architecture, University of Wisconsin–Milwaukee, Milwaukee, Wisconsin 53201

Min Kantrowitz, Min Kantrowitz & Associates, Inc., P.O. Box 792, Albuquerque, New Mexico 87103

David Kernohan, School of Architecture, Victoria University of Wellington, Wellington, New Zealand

Martin Krampen, University of the Arts Berlin, Federal Republic of Germany; Am Hochstrasse 18, D-7900 Ulm-Donau, Federal Republic of Germany

Sonia Kruks, Department of Government, Oberlin College, Oberlin, Ohio 44074

Jon Lang, School of Architecture, University of New South Wales, Kensington, N.S.W., Australia

David Stea, International Center for Built Environment, Santa Fe, New Mexico 87131. *Present address:* Universidad National Autonoma de Mexico, Mexico D.F., Mexico

Martin S. Symes, Department of Architecture, University of Manchester, Manchester M13 9PL, United Kingdom

Ben Wisner, School of Social Science, Hampshire College, Amherst, Massachusetts 01002

Preface

This third volume in *Advances in Environment, Behavior, and Design* follows the conceptual framework adopted in the previous two volumes (see the Preface to Volume 1, 1987). It is organized into five sections—advances in theory, advances in place, user group, and sociobehavioral research, and advances in research utilization.

The authors of this volume represent a wide spectrum of the multidisciplinary environment–behavior and design field including architecture, environmental psychology, facility management, geography, human factors, sociology, and urban design. The volume offers international perspectives from North America (Carole Després from Canada, several authors from the U.S.), Europe (Martin Krampen from Germany, Martin Symes from England), and New Zealand (David Kernohan). More so than any of the previous volumes, they are drawn from both academia and professional practice.

While there continues to be a continuity in format in the series, we are actively exploring new directions that are on the cutting edges of the field and bode well for a more integrated future. This volume will further develop the themes of design and professional practice to complement the earlier emphases on theory, research, and methods.

We have invited leading scholars on design theory to present critical chapters on the comparison and possible integration of explanatory environment–behavior theories and prescriptive design theories. Advances in environment–behavior theory will continue to be a strong focus of this series—but in this volume we are expanding theory to the realm of design theory and the possible integration of environment–behavior *and* design theory.

Since the mid-1980s, we have convened sessions on theory at the annual conferences of EDRA, the Environmental Design Research Association (Atlanta, 1986, Ottawa, 1987), and the biennial conferences of

IAPS, the International Association for the Study of People and their Physical Environments (Haifa, 1986, Delft, 1988). A selection of these papers has been edited and expanded into chapters appearing in this series. In Volume 1, we dealt with a pair of theories diametrically opposed to each other that have led to major debates in the environment–behavior field—empiricism and phenomenology (Volume 1, chapters by Richard Winett and David Seamon). We then moved to two theories that begin to make a bridge between academic research and environmental problem solving—ecological and structural theories (Volume 2, chapters by Gerhard Kaminski and Roderick Lawrence). Now, in Volume 3, we have invited two chapters that explicitly investigate the possible integration of science and design, discipline and profession, and research and practice (chapters by Linda Groat and Carole Després, and Jon Lang).

Fields advance by research focused on the development of theories and by investigations exploring basic conceptual issues. In the multidisciplinary environment–behavior and design field, calls have been made since the early 1970s for the development of theory. In 1973, Amos Rapoport argued for the development of explanatory theory and suggested five reasons why theory was needed: to make sense of findings, to reveal gaps, to lead from description to explanation, to aid in teaching, and to aid in application.

According to a source that seldom does us wrong, the *Oxford English Dictionary, theory* is a "scheme or system of ideas or statements held as an explanation or account of a group of facts or phenomena; a hypothesis that has been confirmed or established by observation or experiment, and is propounded or accepted as accounting for the known facts; a statement of what are held to be general laws, principles, or causes of something known or observed." A *theory*, then, is a set of interrelated concepts held as an explanation for observable phenomena by recourse to unobserved, more abstract principles.

This is the way in which most people have been using the word "theory" in the past 15 or so years of the environmental psychology/environment–behavior field. (See, for example, the work of Irwin Altman [Altman & Rogoff, 1987], Amos Rapoport [1973], and Daniel Stokols [1983].)

It may serve us well to use an adjective before this noun "theory." Thus theory as used in science, by the above authors, and so far in this series is *scientific theory* or *explanatory theory.*

But how about design theories? How do we conceptualize them? Are they also theories? How about the writings of major thinkers in architecture and the other environmental professions like Vitruvius, architect to Caesar Augustus two thousand years ago, who argued that good architecture had three conditions: *utilitas, firmitas,* and *venustas,* or, as we now know them, commodity (function), firmness (technics), and delight (aesthetics)? How about the more recent theories of architecture, like the theory of complexity and contradiction in a book by the same title by Robert Venturi (1966)? How do we conceptualize the writings of Juan-Pablo Bonta (1979) in *Architecture and Its Interpretation* based on structuralism, or of Christian Norberg-Schultz (1980) in *Genius Loci: Toward a Phenomenology of Architecture* based on phenomenology, or of Geoffrey Broadbent (1973) in *Design in Architecture* or Christopher Jencks (1977) in *The Language of Post-Modern architecture* based on the analytic procedures of semiotics? Architectural scholars have always called these "theories," but in what way are they theories? Are they akin to the theories of science, or do they not deserve the term "theory"?

A second type of theory, also covered by the *Oxford*, may lead us out of this apparent dilemma. *Theory*, it suggests, is also a "conception or mental scheme of something to be done, or of the method of doing it; a systematic statement of rules or principles to be followed." Theory in this second sense can be conceived of as a doctrine or ideology, a programmatic idea of how things should be done, almost a manifesto.

In 1984, as a result of a symposium on the topic of doctoral education for architectural research (Moore & Templer, 1984), Peter McCleary proposed a view of "theory" as used in his teachings. Drawn from Bacon's 1626 distinction between *theory* and *practice*, between the speculative and the practical, this view emphasizes that theory can be conceived of as a scheme of ideas that explains practice.

Again, we need an objective to modify "theory." We would agree with our late colleague, Kevin Lynch (his discussion of three normative theories in *Good City Form* [1981]), Jon Lang (in his book on *Creating Architectural Theory* [1987]), and McCleary, that we call this second type of theory *normative theory* or *prescriptive theory* or, in the more limited case, *design theories*. *Design theories*, then, are a scheme of ideas or concepts that relate to observable phenomena and serve, in some way as yet undefined, to "explain" them; but they also serve to define that which is proposed or that which should be done through a program of design action or practice.

To examine these ideas further, we invited Linda Groat and Carole Després to discuss the significance of architectural theory for environment–behavior research, and Jon Lang to explore design theory from an environment–behavior perspective. While there are many points of convergence between these two chapters, the Groat–Després chapter places more emphasis on physical form variables while the Lang chapter takes a more behavioral orientation. Both chapters—individually and as a pair—represent significant new advances to the field.

As in the earlier volumes in the series, we invited another leading scholar, Martin Symes of the University of Manchester, to comment on these two chapters by surveying the relationships between research and design.

The middle part of the volume examines place, user group, and sociobehavioral research topics which have experienced recent empirical developments but which have not yet been covered in the *Advances* series.

Chapters that address place-oriented research in Volumes 1 and 2 of this series varied in context from urban to rural and in scale from sites to regions. They include site-scale environments such as urban plazas, parks, and malls (Mark Francis, Volume 1) and a variation on the theme of urban parks—urban forests (Herbert Schroeder, Volume 2), to regional scales such as rural areas (Frederick Buttel *et al.*, Volume 1), vernacular landscapes (Robert Riley, Volume 2), and developing countries (Graeme Hardie, Volume 2).

The place-oriented research included in this volume fits into the site end of the scale continuum and addresses individual buildings and spaces within buildings. Thomas Hubka provides a new perspective on vernacular architecture, one that expands the scope of the field and that is a companion to Robert Riley's chapter in Volume 2 on vernacular landscapes. Franklin Becker provides a historical review of environment and behavior research in the workplace and discusses key behavioral and spatial issues in the design of facilitative workplaces.

User group chapters in previous volumes have tended to emphasize urban or built environment contexts whether they addressed gender issues (Rebecca Peterson, Volume 1), social groups (William Michelson, Volume 1), or health-care providers in hospitals (Sally Shumaker and Will Pequegnat, Volume 2). Louise Chawla's chapter in this volume follows that same pattern and focuses on children and housing—specifically on policy, planning, and design for housing that recognizes the diversity of family structures within which children now live within the urban context.

Martin Krampen expands on part of the theme of environmental cognition and meaning begun by Tommy Gärling and Reginald Golledge in Volume 2. Initially we had hoped for one chapter that would deal with the entire gamut from perception to meaning, from association to symbolism. But the task was too large. The Gärling–Golledge chapter treated the associational end of the continuum, while Martin Krampen now looks in some detail at three distinctly different traditions for the study of environmental meaning—semiotics, environmental psychology, and ecological psychology. This chapter ought to stimulate much debate and discussion in the field. It merits a follow-up in a subsequent volume in this series, not as a rebuttal, but rather a chapter that might look at the two other major approaches to environmental meaning—the nonverbal communication and the historical-symbolic approaches—and might further develop some of the issues Krampen addresses. To date, one of those issues, Gibson's work on perception, has had only a small impact on the field. We have commissioned a chapter for Volume 4 devoted to the Gibsonian intellectual tradition and its current and potential impacts on our field, not only in terms of meaning but also the more concrete aspects of environmental perception and environmental design.

Considerable attention has been given in the first two volumes of the series to advances in methods for conducting research and utilizing it in environmental problem-solving applications (e.g., chapters on quantitative research methods by Robert Marans and Sherry Ahrentzen and on qualitative methods by Setha Low in Volume 1, on facility programming by Henry Sanoff and postoccupancy evaluation by Richard Wener in Volume 2, and on professional practice by Lynda Schneekloth in Volume 1 and policy development by Francis Ventre in Volume 2). We continue this tradition with two new chapters on research utilization methods in the current volume.

Lynda Schneekloth's chapter in Volume 1 sets the stage for the two current chapters. She made the distinction between information transfer and action research. In a recent dissertation, Min Byung-Ho (1988) extended this distinction philosophically and through case studies to two more general types: a one-community paradigm based on the notion that research and action necessarily go hand-in-hand, and a two-community paradigm based on the notion that research must be independent from and precede action. The two-community approach has the longer and more recognized history in the field of environment and behavior (e.g., the many books of facility programming and postoccupancy evaluation and the series of annual design research awards premiated by *Progressive Architecture*). But the one-community approach is on the ascendancy and is subject to lively debate. We continue to explore its development in the series.

Participatory and action research methods are examined in depth by Ben Wisner, David Stea, and Sonia Kruks. They refer extensively to five detailed case studies in which they have been involved in development contexts, mostly in the Third World. The possible emergence of a new design-decision research paradigm for the field is scrutinized by two leading practitioners, Jay Farbstein and Min Kantrowitz, both principles of major professional firms. The Wisner–Stea–Kruks chapter places emphasis on facilitating participation as a primary goal, whereas the Farbstein–Kantrowitz chapter places emphasis on decision making, even though both chapters are written about a kind of one-community action research. The distinction between participation and decision making may be related to the fact that the former chapter is based on experiences in developing contexts, while the latter is in a developed country.

To round out this section, we invited David Kernohan from New Zealand to comment on emerging research utilization/professional practice methods. He draws a brilliant analogy between environment–behavior research and Einstein's theory in discussing these two chapters and putting them into a larger, more inclusive context.

In the preparation of this volume, we were once again assisted by many people. We would like to thank Ernest Alexander, Friederich Dieckmann, and the students of Theories of Environment and Behavior at the University of Wisconsin-Milwaukee who offered valuable comments on chapter drafts; our Editorial Advisory Board, who offered advice on the series as whole; Michele Lien, who served as editorial assistant for Volume 3; and Eliot Werner, our executive editor at Plenum Press.

ERVIN H. ZUBE
GARY T. MOORE

REFERENCES

Altman, I. (1973). Some perspectives on the study of man–environment phenomena. *Representative Research in Social Psychology, 4,* 109–186.

Bonta, J. P. (1979). *Architecture and its interpretation: A study of expression systems in architecture.* London: Lund Humphries.

Broadbent, G. (1973). *Design in architecture: Architecture and human science.* New York: Wiley.

Jencks, C. (1977). *The language of post-modern architecture.* New York: Rizzoli.

Lang, J. (1987). *Creating architectural theory.* New York: Van Nostrand Reinhold.

Lynch, K. (1981). *Good city form.* Cambridge, MA: MIT Press.

Min, B.-H. (1988). *Research utilization in environment-behavior studies: A case study analysis of the interaction of utilization models, context, and success.* (Doctoral dissertation, University of Wisconsin-Milwaukee.) Ann Arbor, MI: University Microfilms.

Moore, G. T., & Templer, J. (Eds.). (1984). *Doctoral education for architectural research.* Washington, DC: Architectural Research Centers Consortium.

Norberg-Schultz, C. (1980). *Genius loci: Towards a phenomenology of architecture.* New York: Rizzoli. ·

Rapoport, A. (1973). An approach to the construction of man–environment theory. In W. F. E. Preiser (Ed.), *Environmental design research* (Vol. 2). New York: Van Nostrand Reinhold.

Stokols, D. (Ed.) (1983). Theories of environment and behavior. Special issue of *Environment and Behavior, 15* (Whole No. 3).

Venturi, R. (1966). *Complexity and contradiction in architecture.* New York: Museum of Modern Art.

POSTSCRIPT

Effective with this volume, Ervin Zube has resigned as coeditor of the series to devote more time to his research on human landscape transactions and landscape management. Erv and I initiated the idea for the series in late 1983 in discussions with Eliot Werner at Plenum and with the Board of Directors of EDRA (notably Daniel Stokols who suggested the link with Plenum). We have worked together on all aspects of the series since then. Erv's contributions to the series and to the field are immeasurable. His wisdom, guidance, and hard work over the first five years of the *Advances* series will be sorely missed. Fortunately, he has agreed to remain as a member of the Editorial Advisory Board.

Professor Zube will be replaced, starting with Volume 4, by Robert W. Marans, Professor of Architecture and Urban Planning, Director of the Ph.D. Program in Urban Technological and Environmental Planning, and Senior Research Scientist at the Institute of Social Research, University of Michigan. I am delighted to work with Bob Marans on the continuation of the series.

To gain a broader base of advice from North American and European environment–behavior studies, from design and practice, and from the international community, we also welcome Gilles Barbey, Robert Bechtel, Tommy Gärling, Linda Groat, Stephen Kaplan, Setha Low, William Michelson, Rudolf Moos, Toomas Niit, Amos Rapoport, Andrew Seidel, Maija-Leena Setala, Seymour Wapner, and John Zeisel to the Editorial Advisory Board.

Work is progressing well on Volume 4 of the series. Volume 4 will again contain chapters on advances in the domains of the field that have experienced the most rapid development over the intervening years. In preparation are new chapters on urban social theory, Gibsonian ecological theory and cross-cultural theory, environmental aesthetics and the sick-building syndrome, new methodological developments, and research utilization with special attention to housing and facilities for people with Alzheimer's disease and related dementias.

GARY T. MOORE

Contents

2   **Design Theory from an Environment and**   53
                    Behavior Perspective

                           Jon Lang

II ADVANCES IN PLACE RESEARCH

IV ADVANCES IN SOCIOBEHAVIORAL RESEARCH

| 7 | **Environmental Meaning** | 231 |

Martin Krampen

V ADVANCES IN RESEARCH UTILIZATION

Ben Wisner, David Stea, and Sonia Kruks

Jay Farbstein and Min Kantrowitz

10   **"Einstein's Theory" of Environment–Behavior**   **319**
                 Research: A Commentary on Research Utilization

                     David Kernohan

Advances in Environment, Behavior, and Design
VOLUME 3

I

ADVANCES IN THEORY

The Significance of Architectural Theory for Environmental Design Research

LINDA N. GROAT and CAROLE DESPRÉS

The main thesis of this chapter is that the environment–behavior research must, if it is to be seen as relevant to architectural design, incorporate not only empirically based theory derived from the social sciences but architectural theory as well. The intention of the chapter is to explore the nature of architectural theory and how it can be usefully integrated in environment–behavior research.[1] This thesis is supported by a sequence of three interrelated arguments.

First, environment–behavior research, by definition, must fundamentally concern itself with the description and explanation of the key physical attributes which are significant in people's experience of the

[1]The term "environment–behavior" is being used in order to maintain consistency with the terminology used throughout this book. However, the authors ordinarily prefer the term "environmental design research" since it not only avoids any implications of the subject–object duality but also suggests the necessity of building upon the literature associated with design-based disciplines.

Linda N. Groat • College of Architecture and Urban Planning, University of Michigan, Ann Arbor, Michigan 48109.   **Carole Després** • School of Architecture, Laval University, Quebec G1K 7P4, Canada.

built environment. Although this argument may seem self-evident, a number of researchers have, in fact, pointed out that much of the current environment–behavior research has dealt inadequately with issues of physical form.

Second, it is further argued that the body of literature usually termed "architectural theory" can provide an appropriate basis for studying the significant physical attributes of the built environment. The conventional wisdom among many environment–behavior researchers is that architectural theory is no theory at all; or, if it is theory, architectural theory must certainly not be considered relevant to the same domain of inquiry. It is the contention of this chapter that historically there has been more of a relationship between the two than is currently apparent or acknowledged. In this light, it is argued that the perceived discontinuity between the domain of architectural theory and that of environment–behavior research may, in fact, be simply a temporary ontological and epistemological disjuncture. Furthermore, it is argued that it is precisely by trying to identify the nature of the relationship between the two bodies of theories that significant advances may be made in environment–behavior research.

And third, this chapter demonstrates the potential significance of this approach by identifying a number of relevant examples of both theoretical analyses as well as specific environment–behavior research studies.

GETTING PHYSICAL: THE NEED TO IDENTIFY FORMAL ATTRIBUTES

Since the mid-1970s, a number of environment–behavior researchers have pointed to the lack of specificity with which physical form is both described and analyzed in much of the current research in the field (e.g., Canter, 1977; Groat, 1983; Sime, 1986; Wohlwill, 1976). Canter (1977), in particular, has commented on the tendency of much environment–behavior research to minimize the role of physical form. He suggests that "the specification of the physical constituents . . . is a much more significant component [of built form] than the research literature would have one believe" (p. 159). He then speculates about why this is so:

> The reason for such a dearth of studies appears to be, in part at least, the difficulty of deciding which physical attributes to study. Taken in the abstract, independently of a conceptual framework, there are an infinity of ways of dividing up and measuring physical parameters. Weight, size, col-

our, shape, . . . and many others, at any scale are feasible. So researchers
have either selected one which caught their fancy, with disappointing results, or given up because they were spoilt for choice. (Canter, 1977, p. 159)

More specifically, in the area of environmental aesthetics—a topic
which by definition should address the physicality of the environment—Wohlwill (1976) has argued that the information derived from
dimensional analysis of semantic structure is simply descriptive of the
way in which respondents use the adjective scales "without telling us
anything about the role played in these judgments by any specific environmental characteristic" (pp. 60–61). More recently, in reviewing the
concept of "place" in the environment–behavior literature, Sime (1986)
concludes that many researchers continue to neglect "the physical context of behavior and experiences" (p. 49).

Ironically, as Groat (1983) has already pointed out, there is actually a
vast literature in architectural theory that provides just the sort of "conceptual framework" for "dividing up and measuring physical parameters" that Canter suggests is necessary. And yet this wealth of material
on the description of physical form seems to have been largely ignored,
dismissed, or misunderstood by all too many environment–behavior
researchers. Given that many researchers have been trained exclusively
in nondesign disciplines, this state of affairs, though regrettable, may
not be surprising. More worrying, however, is that this lack of interest in
architectural theory seems at least partially shared by a number of design researchers with architectural training.

In a somewhat different vein, Jon Lang's book *Creating Architectural
Theory* (1987) makes an important contribution to the field by examining
the variety of ways in which environment-behavior research can contribute to the more rigorous development of architectural theory. Yet, he
seems to consider only minimally the possibility of theoretical contributions being made the other way round, i.e., architectural theory contributing to environment–behavior research. In describing the nature of
architectural theory, he defines architectural theory as "normative,"
meaning that it is descriptive and prescriptive in nature; it "consists of
statements on what ought to be—what a good world is" (p. 15). He then
adds: "These assertions are based largely on the insights and personal
experiences of the individual professional. . . . However observant the
designer may be, the conclusions he or she draws about the way the
world works are biased by that individual's contacts with the world" (p.
16).

Other characterizations of architectural theory seem to focus on its
apparently contentious and polemic qualities. Based on discussions with
several colleagues in both environment–behavior research and architec-

tural criticism, Robinson (1986) concluded, "The present view of architectural theory seems to be that it consists of a variety of competing and opposing views of architecture which are in opposition to each other" (p. 98), a view which Robinson herself sought to challenge.

Unfortunately, the perspective Robinson cites tends to minimize the extent to which architectural theory can reflect valuable insights about how physical form may be interpreted, particularly its relationship to both historical and contemporary design principles. It is our contention, however, that discourse in architectural theory does, in fact, offer evidence of a body of shared knowledge that can contribute to the advancement of environment–behavior research. To appreciate this potential, however, requires a more elaborate discussion of the nature of architectural theory.

THE NATURE OF THEORY IN ARCHITECTURE AND ITS LINKS TO SCIENTIFIC EPISTEMOLOGY: AN HISTORICAL PERSPECTIVE

In order to clarify the nature of architectural theory and its relation to environment–behavior research, it is necessary to broaden the discussion beyond the realm of the two fields as they are presently constituted. This requires: (1) an historical account of the development of architectural theory and (2) an analysis of the broader cultural milieu of which architectural theory and the social sciences are both inevitably a part.

Such an undertaking is, of course, not without some potential pitfalls. The most obvious of these is the unavoidable reductionism that necessarily accompanies any attempt to summarize broad historical and cultural trends, especially if it involves differentiating among historic "periods." Some contemporary theorists would even argue against any attempts to describe history in this way. Nevertheless, the contemporary cultural historian Hassan (1987) argues that it is, indeed, useful to characterize historic periods in order to combine chronological and cultural history. To minimize the problem of reductionism, he proposes that the concept of the historic period be treated as both a diachronic (historical) and a synchronic (theoretical) construct, embodying not only continuity but discontinuity as well. In other words, it is important that both the essential continuities and the more disjunctive moments be seen as complementary forces.

With these caveats in mind, Table 1 represents a summarized history of architectural theory from the Renaissance to the present. In this context, it is useful to consider the evolution of architectural theory in

TABLE 1. An Historical Perspective on the Nature of Architectural Theory

	Renaissance-baroque 1450–1700	Premodern 1750–1880	Modern 1910–1960	Postmodern 1965–
Relation to cultural milieu	Shared elite culture	Disputed culture	New "mass" culture	Fragmented culture
Rhetoric	Beauty	Authenticity	Functional meaning	History and meaning
Design principles	Hierarchical ordering	Independence of design elements within traditional styles	Asymmetry and dissonance	Simulacre, collage, and decomposition
Relation to scientific epistemology	Validation through cultural convention	Rationalization of architectural procedures	Adaptation of positivist approach to science	Post-positivist critique

terms of four periods: Renaissance–baroque, premodern, modern, and postmodern. Each of these periods is described in terms of four issues: (1) *relation to cultural milieu,* which describes the broad sociocultural backdrop for each era, particularly the relationship among the various segments of society influencing architectural production; (2) *rhetoric,* which identifies the ideals and theoretical concepts that were most central to the architectural discourse of the time; (3) *design principles,* a term that identifies significant aesthetic norms, particularly as they apply to the actual composition of built form; and, (4) *the relation to scientific epistemology,* which describes the ways in which discourse on architectural theory has incorporated significant concepts associated with scientific thought in each era.

Thus, the following historical overview has two related purposes. First, by highlighting the gradual shifts in the substantive focus of theoretical discourse in architecture, it challenges some of the rather limited interpretations of architectural theory in the current environment–behavior research literature. And second, by framing the discussion within broader sociocultural trends, this analysis suggests that architectural theory has, indeed, been more fundamentally linked to the development of scientific epistemology (of which the emerging social sciences were a part) than has generally been acknowledged.

THE RENAISSANCE–BAROQUE TRADITION

Relation to cultural milieu. The term Renaissance–baroque describes an architectural tradition that spans the beginning of the Renaissance in the fifteenth century to the end of the baroque era in the early eighteenth century. This architectural tradition emerged and eventually prevailed within the context of a larger milieu that was essentially a shared elite culture. The word *shared,* however, requires some elaboration because it is used in a rather limited sense to describe two distinct historical phenomena.

First, the cultural ideals of the Renaissance–baroque tradition were shared primarily among the elite social strata within the context of "centralized authoritarian systems" (Norberg-Shultz, 1971, p. 258). With respect to architectural production, architects and their patrons (typically nobility or churchmen) appear to have held common beliefs about the nature of architecture and its relation to both society and universal order. This situation is made explicit in Alberti's *Ten Books of Architecture* (1452/1988), the first major treatise on architecture since Vitruvius' text of the first century B.C. As Kostof (1985) has pointed out, Alberti "wrote not so much as a practitioner speaking to other practitioners, but as a humanist explaining to the important and rich people of his day about the exalted profession of architecture and its place in public life" (p. 407). Indeed, Alberti advises other architects to "deal only with principal citizens who are generous patrons and enthusiasts of such matters: a work will be devalued by a client who does not have an honorable situation" (p. 318).

Second, the cultural milieu of the Renaissance–baroque period can be considered a "shared" one in yet another sense. Despite the turmoil generated in the Reformation–Counter-Reformation and devastating wars among rival nations, this artistic–architectural tradition was embraced everywhere in Europe by the early 1600s and continued to be maintained through the seventeenth century (Kostof, 1985, p. 511).

Rhetoric.   The essence of the Renaissance–baroque ideal of architectural form was *beauty* (e.g., Alberti, 1988; Kaufmann, 1955; Lesnikowski, 1987; Vidler, 1987). To a great extent, this ideal derived from the Renaissance revival of the earlier Vitruvian text as well as careful analysis of Roman monuments. In choosing to "rewrite" Vitruvius (Murray, 1978), Alberti clearly reaffirmed the fundamental importance of classical proportions in achieving beauty when he argued that every building element must be "reduced to exact measure, so that all parts will correspond to one another" (as quoted in Kaufmann, 1955, p. 92). Moreover, qualities such as proportional harmony and quantitative perfection were considered manifestations of universal law (Kaufmann, 1955; Kostof, 1985).

Design principles.   Although the ancient classical ideal of quantitative perfection was revived in the Renaissance, the design principles espoused in Renaissance–baroque theory nevertheless represented a new design system. More specifically, Renaissance–baroque architecture represented a shift from the classical emphasis on proportions for their own sake to a new sensibility whereby the hierarchical ordering of the whole was seen as the primary goal.

This shift in design systems can be explained by comparing the organization of a Greek temple with that of a typical Renaissance building. Generally speaking, the various elements of a Greek temple are of equal value; columns, for instance, are all of the same height. In contrast, the various component parts of a Renaissance building are differentiated so that certain elements are subordinate to others; in other words, some sets of columns may, in fact, be smaller than others. The result is that the whole is conceived as a "hierarchy of well-disciplined elements" (Kaufmann, 1955, p. 79).

Despite the various stylistic fluctuations which occurred during the period from 1450 to 1700—particularly Mannerism—the basic system of

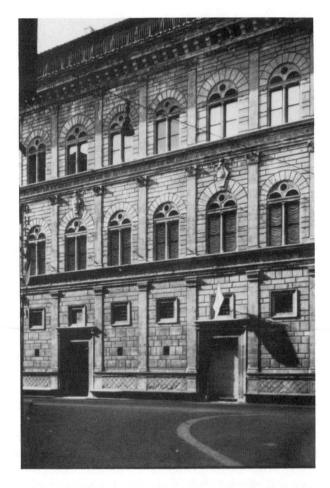

FIGURE 1. Renaissance architecture is typified by the Palazzo Rucellai by Rosselino (from the design by Alberti), Florence, Italy, 1446-1451. Courtesy of the Art & Architecture Library, University of Michigan.

hierarchical ordering was nevertheless maintained (Kaufmann, 1955). Although this system was sometimes reinterpreted, exaggerated, and otherwise challenged during the various stylistic eras (including Mannerism and later baroque), it was not actually violated until the emergence of premodern architecture. Indeed, "[m]any of the treatises of the later Baroque . . . prove that the architects of the seventeenth and eighteenth centuries . . ." clung to the doctrine of the proportions (Kaufmann, 1955, p. 84).

Figure 2. A notable example of Mannerist architecture is the Romano House by G. Romano, Mantua, Italy, 1544. Courtesy of the Art & Architecture Library, University of Michigan.

Relation to scientific epistemology. Architectural theory in the Renaissance–baroque period was inherently linked to the Renaissance search for a natural and universal order. Since the cosmos was conceived in terms of absolute numbers, architecture was therefore considered "a mathematical science which ought to make the cosmic order visible" (Norberg-Shultz, 1980c, p. 113). In this light, then, it is perhaps not surprising that Alberti claimed: "Architecture is a very noble science" (Alberti as quoted by Kostof, 1985, p. 407). Kostof (1985) elaborates:

> The architect, armed with the science of linear perspective and the new mathematics, steeped in the knowledge of ancient sources, becomes the master of a universal law that applies as much to the frame of his buildings as it does to the structure of the natural world. And since nature in Alberti's thought is synonymous with God, the architect in his pursuit approaches the divine. (p. 408)

By the end of the seventeenth century, however, the unity of art and science began to dissolve. The "artist no longer dared to be a philosopher or scientist, and as a consequence artistic theory lost much of its impetus during the seventeenth century" (Norberg-Shultz, 1971, p. 12).

Figure 3. Baroque architecture is exemplified by the facade of S. Carlo alle Quattro Fontane by Borromini, Rome, Italy, 1665–1667. Courtesy of the Art & Architecture Library, University of Michigan.

And in a similar vein, Perez-Gomez (1983) identifies trends in the seventeenth century that would lead eventually to a questioning of the link between the human and the divine, thereby calling into question the role of geometrical order in architecture.

THE PREMODERN TRADITION

Relation to cultural milieu. The term "premodern" describes the period that begins with the gradual dissolution of the Renaissance–

baroque tradition at the end of the eighteenth century and continues to the end of the nineteenth century. The cultural milieu of this period represents a significant challenge to the *a priori* axioms of Renaissance–baroque thought and is manifested in the subsequent emergence of an increasingly relativist stance. It is in this sense that the cultural milieu of this period can be described as a disputed one. Perez-Gomez (1983) has characterized the overall transformation of the cultural ethos during this period as "the divorce of faith from reason" (p. x). According to Perez-Gomez, the material world, including architecture, gradually came to be seen as inanimate, that is, without metaphysical significance.

Among the more significant sociocultural trends affecting architectural theory were institutional reform and historiography (Vidler, 1987). With regard to the former, new ideas in social philosophy, economic theory, medicine, and the physical sciences generated important debates regarding the appropriate nature of a variety of institutions including: factories, hospitals, prisons, and workhouses. These debates revealed important differences of opinion about established sociocultural conventions. Second, the gradual development of research in history and anthropology eventually touched off a sense of cultural relativism. Both ancient and distant cultures needed to be explained in their own terms rather than in reference to the *a priori* axioms of the previous era.

Rhetoric.   The theoretical emphasis in premodern architectural discourse gradually shifted from the ideals of beauty to a search for the origins of architecture (Vidler, 1977, 1987). This shift was consistent with the emerging relativism of the era and was manifested in a tendency to "return all subjects to their 'natural' and therefore 'principled' origins" (Vidler, 1987, p. 3). Whereas the ideal of beauty in the Renaissance–baroque tradition had been understood to be real and absolute (Honour, 1968), the search for the origins of architecture betrayed an uncertainty about the theoretical and ethical basis of the discipline. It is in this sense that the term *authenticity* best describes the theoretical focus of most premodern theory.

The significance of this theoretical focus is manifested in a variety of treatises. One of the most influential was Laugier's essay on the primitive hut, published in 1753 (Laugier, 1977). In it he argued that the "hut" was the root of all architecture. "From it he [Laugier] derived the essential elements of architecture . . . , in the same way that Rousseau two years before had set up a model of 'natural' man by means of which to criticize contemporary civilization" (Vidler, 1977, p. 95). Subsequently, some of the theoretical principles underlying Laugier's argument were further elaborated by the introduction of the concept of "type"; by the latter part of the century, the term *type* was already prevalent in discourse on the arts. Among the more significant treatises on this topic was one by Quatremere de Quincy (1832), who argued that the hut, together with the cave and the tent, constituted three types or founding models for architecture. Of these, however, only the hut—the type for classical architecture—was amenable to generalization for the production of "all civilized architecture. . . . In this sense, type acted to explain regional and cultural differences while at the same time asserting a fixed and preferred standard, a frozen classification of an otherwise endlessly relativized history" (Vidler, 1987, p. 151).

Design principles. Since classicism remained the primary vocabulary of architecture until well into the 19th century, it might at first appear that the fundamental principles of the Renaissance–baroque system had remained intact. This appears, however, not to be the case. Instead, in the eighteenth century there gradually emerged a tendency to turn away from the principles of compositional hierarchy and to emphasize individual building elements apart form the proportions and harmony of the whole (Kaufmann, 1955). In that sense, then, premodern architecture represented a fundamentally new design system which was manifested in at least three important ways. First, the Renaissance–baroque ideal of geometrical perfection was gradually "subordinated to the idea of a geometrical order that followed the dictates of social and environmental needs," a tendency which in an important sense constituted a kind of "primitive functionalism" (Vidler, 1987, p. 3). Second, the concept of type was further elaborated by the notion of character, the means by which each building of a type would be individualized and differentiated from every other. The result was that a number of eighteenth-century architects, including Boulee and Ledoux, generated "an endless play of abstract geometrical permutations" that ultimately dissipated the integrity of the type (Vidler, 1977, p. 103). And third, philosophical concepts, such as Locke's theory of sensations and Burke's notion of the sublime, led to a concern for the achievement of "the legible facade." As a consequence, architects sought to make the purpose of buildings, particularly those of new or "non-architectural building types" absolutely legible (Vidler, 1987, p. 3).

Relation to scientific epistemology. To a large extent, the increasingly relativistic stance that emerged at the beginning of the premodern era is directly linked to the development of science. The hallmark of this epistemological revolution was positivism, which can be characterized as a belief in "the infinite capacity of human reason" (Perez-Gomez, 1983). In other words, an understanding of the world—based primarily on the power of reason—replaced the Renaissance–baroque belief in a world "deduced from a few immutable a priori axioms" (Norberg-Shultz,

FIGURE 4. The tendency to turn away from principles of compositional hierarchy and to emphasize individual building elements is strikingly manifested by the gradual change in the designs of Sir John Nash's housing terraces in London. Here the geometric volumes of Cumberland Terrace (1826–1827) contrast with the more conventionally ordered hierarchy of the earlier Chester Terrace (1825). Courtesy of the Art & Architecture Library, University of Michigan.

1980b, p. 7). Reason, now harnessed in the service of a new empiricism, was applied to the observation and study of phenomena themselves rather than to the deduction of "facts" from *a priori* assumptions.

The emergence of a scientific epistemology was closely linked to many of the concepts underlying premodern architectural discourse. For example, Laugier's notion of the primitive hut can be understood as a manifestation not only of Rousseau's philosophy but also of the belief in "single causes and their rational consequences, according to a coherent model, rather like that provided by Newton for the physical sciences" (Vidler, 1987, p. 3). Similarly, the concept of "type" owes much to the work of such natural scientists as Buffon, Adamson, and Linnaeus, who sought to identify a prototype for each species (Vidler, 1977, 1987). And finally, the premodern preoccupation with authenticity— that is, with the essential origins of architecture—was substantially informed by the emerging scientific and social science disciplines, such as natural history and anthropology (Vidler, 1987).

THE MODERN MOVEMENT

Relation to cultural milieu. The modern movement in architecture emerged at the close of the nineteenth century, came to full flower in the heroic era of modernism in the 1920s, flourished in international influence after World War II, and ultimately found its preeminence challenged in the 1960s. On a philosophical level, modern thought has been characterized as "a cultural template of liberation from traditional ways of thinking, believing, and acting—and later from traditional itself" (Rochberg-Halton, 1986, p. x). Similarly, the ideals of the modern movement in architecture emerged gradually as a liberation from the sociocultural inequities and constraints characteristic of the nineteenth century. Chief among the trends that influenced the modern movement architects were the force of the scientific and technological advances associated with industrialization, the impact of the Darwinian concept of evolution and progress, and a recognition of the vast sociocultural inequities which emerged as a consequence of the industrial revolution (Brolin, 1976).

Within this broader context, architectural theory was characterized by a simultaneous faith in "the liberating aspects of industrialization" (Jencks, 1986, p. 27) and the virtues of mass democracy. The most influential architects (and artists) of the era saw themselves as the avantgarde leaders of a new mass culture; they sought both to establish a new society (based on the achievements of science and technology) and to generate an architecture of the times from first principles, a process which would presumably eliminate both the physical and social ills of the past.

Rhetoric. The rhetoric of the modern movement discourse was dominated by a concern for functional expression. The rallying cries of "the house is a machine for living in" (Le Corbusier, 1960, p. 10) and "form follows function" are the most frequently cited and obvious examples of this rhetorical stance. Emphasizing the significance of these slogans, however, tends to mask the multifaceted ways in which functional expression came to be understood within architectural discourse. Newman (1980) identified two distinct strains of functional expression within the modern movement. He argues that, after a period of essentially unified rhetoric in the 1920s, an intellectual schism began to develop between what he labels as the "style metaphysicists" and the "social methodologists." The former tended to emphasize the formal and aesthetic metaphors used to express modern technology, whereas the latter group tended to emphasize logical design processes and the social purposes of the new architecture. These tendencies vied with each other for influence in the movement; and, at least according to Newman, it was the style metaphysicists who ultimately prevailed.

FIGURE 5. The "style metaphysicist" approach to modern architecture is represented by Chicago Alumni Memorial Hall by Mies Van der Rohe (1945–1946). Courtesy of the Art & Architecture Library, University of Michigan.

Figure 6. The "social methodologist" approach to modern architecture is notable for its intentional responsiveness to multiple human and functional criteria. It is manifested in the variety of formal elements in Alvar Aalto's design for the library at Mt. Angel Benedictine College, Salem, Oregon, 1970. Photographs by Linda Groat.

FIGURE 6 (cont.)

Design principles. Although the basis for the design principles of modernist architecture are alluded to in any number of texts, Zevi's book *The Modern Language of Architecture* (1978) represents one of the most explicit accounts. Zevi's intent is to present a coherent codification of modernist architecture that matches the clarity with which the classical code has been articulated. In doing so, he offers a set of seven "invariables" that he believes constitute the code of modern architecture. For example, the invariable of asymmetry and dissonance represents a direct disavowal of the classical principles of symmetry and order. Another, antiperspective three dimensionality, calls for the asymmetrical disposition of buildings in relation to each other. A third, the syntax of four-dimensional decomposition, defines the disintegration of the volume so as to emphasize the expression of planes. Taken together, Zevi's seven "invariables" demonstrate the extent to which modernist architects rejected the previously established design principles. Although Zevi's characterization of these principles as "invariables" could certainly be disputed, they nevertheless represent some of the aesthetic concepts most typical of modernist architecture.

A complementary perspective on the aesthetic principles of modernist architecture is presented in a recent text by Herdeg (1983).

Herdeg's particular contribution is to focus on the epistemological implications of the Bauhaus design principles as they were promulgated by Gropius. Although Herdeg deals exclusively with an analysis of the Harvard architecture program (which Gropius headed between 1937 and 1953), the profound influence of this design program throughout the United States and elsewhere makes this an important study. According to Herdeg's analysis, the two essential characteristics of Bauhaus-legacy design are (1) a diagrammatic plan, conceived as "a literal expression of functional relationships" (p. 2), and (2) an exterior design based on clever variations in patterning to achieve visual stimulation. The former not only represents the outcome of the "objective" analysis of program but, more importantly, it also reflects the assumption that careful analysis would lead inevitably to the appropriate architectural form in a kind of cause-and-effect relationship. The second principle, visual patterning, reflects Gropius's belief that aesthetic forms could become "neutral carriers" of abstract meaning, capable of expressing whatever the designer wanted to express (p. 94).

Relation to scientific epistemology.   One of the consequences of the development of modern thought is, according to Rochberg-Halton (1986), a culture of nominalism, in which the world is understood as "based either on conventions themselves ultimately arbitrary" or "upon 'untouchable' natural laws reflecting a mechanical order of things" (p. 234). The former tendency, which Rochberg-Halton labels *mythic relativism,* is manifested in the belief that we experience "an infinitely plastic world, a world largely . . . arbitrary at base" (p. 235). The latter tendency, *mythic objectivism,* is manifested in the belief that "all nature is based on mechanical law . . . outside the sphere of human modification" (p. 234). The ongoing tension between these two tendencies, he suggests, represents "the central cultural opposition of our times" (p. 235).

This cultural opposition appears to be reflected in the "subjective/objective seesaw" which, according to Herdeg (1983), characterized the Bauhaus legacy. To use Rochberg-Halton's terms, mythic relativism is clearly manifested in Gropius's (and others') belief that the manipulation of visual patterning could achieve an aesthetic quality that would mean what the designer wanted it to mean. On the other hand, mythic objectivism is evident in the modernists' adoption of scientific concepts. As the reproduction of the illustration titled "The Plan Calculates Itself from the Following Factors" (produced by the Bauhaus in 1930) suggests, the assumption was that precise programmatic analysis would mirror the sort of cause–effect relationships that were thought to be integral to an "objective" science. In a similar vein, Lang (1987) has traced the influences of empiricism—particularly in the field of experimental psychology—on the aesthetic principles on modernism; and Senkevitch

(1983) has provided a detailed account of the influence of perceptual psychology on the Russian rationalists' attempt to formulate "an objective framework of formal principles for organizing the elements of their new design vocabulary into a dynamic architectural whole" (p. 80).

POSTMODERNISM

Relation to cultural milieu. Any attempt to characterize the postmodern condition is inevitably analogous to shooting at a moving target from a nonstationary vantage point. Not only do we not know how long this entity—postmodernism—will remain within our field of vision, but we are inevitably participants in the process by which it evolves. These are, of course, the unavoidable limitations of any commentary on contemporary society. No wonder then that the twentieth-century cultural historian Hassan (1987) has commented that there is "no clear consensus" about the meaning of the term "postmodernism" (p. 87). He further acknowledges that, at least at the moment, there is a "clearer consensus on the modernist break with *its* tradition" than there is on the postmodernist break with the modern tradition (p. 214, Hassan's emphasis).

Despite these limitations, however, it is still possible to summarize some of the most salient characteristics of postmodernism. This new sensibility is marked by a shift "from a relatively integrated mass-culture to many fragmented taste cultures" (Jencks, 1986, p. 43). Whereas modern culture was based on the "endless repetition of a few products," the postmodern culture is based on short-run production and "the targeting of many, different products" (p. 48). This view of postmodernism in the broad cultural context is paralleled in the realm of architecture. Jencks (1986), in particular, has suggested that postmodern architecture "is a movement that starts *roughly in 1960 as a set of plural departures from Modernism*" (p. 22, Jencks's emphasis). And although there still remains some disagreement as to the exact range and nature of that "pluralism" (e.g., Davis, 1978, vs. Jencks, 1977), postmodernist culture is nevertheless commonly characterized as pluralistic and fragmented.

Rhetoric. One consequence of this pluralism is that the key theoretical concepts of postmodern architectural theory are to some extent disputed. Although the notions of both *meaning* and *history* are central to postmodern discourse, there are at least two distinct perspectives on how these concepts are relevant to contemporary architecture. According to the first view (Foster, 1984), two distinct strains of postmodernism can be identified. The "neoconservative" strain in both art and architecture "is marked by an eclectic historicism in which older and newer modes and styles . . . are retooled and recycled" (p. 146). This reclamation of history is substantially motivated by the desire to reestablish a basis for meaning in architecture, such that built form might once again serve as a medium of shared cultural expression (e.g., Broadbent, 1977; Jencks, 1977). In contrast, the poststructuralist position (which loosely derives from deconstructionist literary criticism) assumes both the fragmentation of history and the impossibility of any "objective" subject matter in art and architecture (Foster, 1984). Its intention is "to reveal multiple, contradictory, and irreconcilable meanings" (Horn, 1988, p. 41), thereby rendering all cultural meanings "ambiguous" and "indeterminate" (Foster, 1984, p. 152). Foster (1984) argues, however, that despite the superficial differences between the two strains of postmodernism, the net effect of both is that the sense of history is eroded and the traditional language of architecture is destroyed.

Not all contemporary critics subscribe to Foster's analysis. According to the second view of postmodernism (Jencks, 1986), much of what has been described by others as postmodernism—including Foster's poststructuralist postmodernism—is in reality "late modernism." The term "late modernism" is taken to mean an architecture which "from about 1960 takes may of the stylistic ideas and values of modernism to an extreme in order to resuscitate a dull (or cliched) language" (p. 32). Similarly, Vincent Scully has described the poststructuralist tendency in architecture as the "terminal twitch of modernism" (Horn, 1988, p. 40). In his view, it is the "neoconservative" strain that represents the true postmodernism. Moreover, in contrast to Foster, Jencks argues that postmodernism can, indeed, reclaim history and meaning as real and positive forces in architecture. In any case, postmodernism in architecture—according to nearly any definition of the term—is generally characterized by either an affirmation or radical denial of history and meaning as important forces in architectural production.

Design principles.   Given the extent to which pluralism defines the postmodern condition, it almost seems a contradiction in terms to try to identify common design principles. Indeed, Jencks (1986) has argued that there is no one postmodern style. Nevertheless, the notions of collage, simulacre, and decomposition represent three of the most salient features of contemporary architecture.

The concept of *collage* generally describes the tendency in much postmodern architecture toward eclectic historicism. Various elements from one or more historic styles—including modernism—may be combined in a single composition. However, as both Jencks (1986) and Eco (1984) have suggested, such quotations of the past are typically assembled together with an intentional irony. A second related principle is that of *simulacre*, the semblance of something else. This notion is man-

FIGURE 7. This housing project in London by Jeremy Dixon (1976–1979) expresses the quality of *collage*, which is typical of the neoconservative strain of postmodernism. Photograph by Linda Groat.

ifested in the straight stylistic revivalism of architects such as Quinlan Terry and Allen Greenberg. And third, the concept of *decomposition* implies the disintegration of the traditional architectural language, an intention which is overtly expressed in poststructuralist architecture. The "architectural certainty" of wall, floor, and ceiling planes are no longer taken for granted (Giovannini, 1988).

Relation to scientific epistemology. In its broadest possible terms, postmodernist thought represents a direct critique of scientific epis-

FIGURE 8. The principle of *simulacre* can be seen in the design of the St. Matthew's Episco-
pal Church in Pacific Palisades, California (1979–1983), by Charles Moore and his partners
Ruble and Yudell. Here the architects' goal was to evoke "the semblance of something
else"—i.e., the original, woodsy, Bay Region church which had burned down. Photograph
by Linda Groat.

temology. Among philosophers, for example, the term "postmodern"
refers to virtually all post-positivist thinkers who share in common a
rejection of modern logical positivism (Jencks, 1986). And in architecture, postmodernist theory appears fundamentally to reflect a "crisis of
confidence in science and technology" (Horn, 1988, p. 42).

What remains in dispute among contemporary theorists, however,
is the particular philosophical stance that offers the most viable way
forward beyond the "materialistic and mechanistic" (Ittelson, 1989) assumptions inherent in current scientific theories. For example, Colquhoun (1988) has suggested that structuralism, and especially as it is
manifested in semiological analysis, has provided an important theoretical basis for the postmodern critique of modern functionalism. Although Colquhoun also acknowledges the limitations of such analyses,
the influence of this perspective has been highly influential, especially
in the earliest examples of postmodern theoretical discourse in architecture (e.g., Broadbent, 1977; Jencks, 1977).

A second critique of modernist epistemological assumptions in con-

FIGURE 9. The principle of *decomposition*, typical of poststructuralist architecture, is expressed in the design of the architect Frank Gehry's own house in Santa Monica, California (1978). Photograph by Linda Groat.

temporary theory is provided by the phenomenological perspective. Perez-Gomez (1983) is perhaps the most explicit in his critique of contemporary architectural discourse. He argues that the transcendental aspects of architecture (and architectural theory) came gradually to be rejected as a consequence of the emergence of modern science and its preoccupation with the internal logic of the objective systems of built form. In light of this trend, Perez-Gomez argues that the phenomenological perspective offers the appropriate theoretical foundation for advancing architectural theory beyond the crisis engendered by the development of modern science and modernism.

Finally, some architectural theorists have begun to argue that *both* the structuralist and the phenomenological critiques of modernism are still mired in the epistemological constraints of modern thought and are, therefore, incapable of providing a comprehensive and robust basis for architectural theory. For example, Colquhoun (1988) suggests that the structuralist perspective is too often linked to the presumed autonomy of the linguistic system. Similarly, Dovey (1988) has recently criticized the phenomenological perspective for its insistence on "the autonomy of the subject" (p. 278) and its lack of attention of social structure and ideology. In effect, then, both Colquhoun and Dovey have presented two complementary analyses of the apparent limitations of the theoretical bases for recent postmodern critiques of the modern tradition in architecture.

THE RELATION BETWEEN ENVIRONMENT–BEHAVIOR RESEARCH AND ARCHITECTURAL THEORY: RAPPROCHEMENT OR DISJUNCTURE?

The preceding historical overview has been presented with two purposes in mind. First, the intention has been to highlight the gradual shifts in architectural theory over time and thereby challenge the often static, limited interpretations of architectural theory presented in some of the environment–behavior literature. And second, the intention has been to demonstrate the extent to which the theoretical concepts central to architectural discourse have been linked to emerging trends in science. Taken together, these observations begin to suggest at least three reasons for the commonly held belief among researchers that architectural theory is essentially unrelated to the domain of environment-behavior research theory.

The first source of misunderstanding, we would argue, is a consequence of the conflation of two distinct domains of architecture: the profession and the discipline. In a recent paper, Anderson (1988) describes the distinction between the two. He defines the *profession* as being "concerned with the current structure of practice." It is mainly synchronic and synthetic in focus; it is "also inherently projective—it brings something into being." In contrast, he defines the *discipline* as "a collective body of knowledge that is unique to architecture and which, though it grows over time, is not delimited in time or space."

Although these two domains intersect and may make contributions to each other, there are aspects of each that are not shared. For example, issues concerning office management and marketing of professional services fall largely within the domain of the profession. On the other hand, examples of eighteenth- and nineteenth-century discourse that debate the significance of proportions as the primary basis for successful architecture belong primarily to the domain of the discipline.

In order to elaborate this argument, however, it is necessary to describe the prevailing assumptions about the nature of *theory* in both the architectural and environment–behavior literature. In a recent paper, Moore (1988) suggests that a major difference between the two bodies of theory concerns the "type" of theory they embody. Specifically, he suggests that a scientific [design research] theory is "a set of interrelated concepts held to be an explanation for observable phenomena by recourse to unobserved, more abstract principles"; in contrast, a design theory relates to observable phenomena and serves *"in some way as yet undefined"* (our emphasis) to explain them. But more importantly, a design theory also serves "to define that which is proposed, or . . . that which should be done through a program of design action." In other words, scientific and design theories are held to be two different types of theories, the former conceived as *explanatory* theory, the latter as *prescriptive* or *normative* theory.[2]

In this regard, Moore's discussion and terminology echoes those of other theorists and researchers (e.g., Lang, 1987). However, Moore's argument that environment–behavior theory can be equated with explanatory theory is only valid if one accepts a relatively narrow definition of what constitutes environment–behavior research. And similarly, his description of architectural theory as essentially normative is valid only if one considers a likewise narrow definition of architectural theory.

In contrast to this view, we suggest instead that both architecture and the design research fields make use of both explanatory and normative theory. This alternative perspective can be more fully articulated by describing a second import issue, testability. Moore suggests that the principle of testability represents *the* critical demarcation between scientific (environment–behavior theory) and architectural theory. As many philosophers of science have argued (e.g., Popper, 1965), a necessary characteristic of any scientific theory is that it be at least testable in principle. The crux of Moore's argument is that explanatory theories—unlike normative (architectural) theories—"are testable in relation to empirical reality."

It is our contention, however, that this view of architectural theory is too limited and, therefore, inaccurate. To be sure, many proposals for design action found in architectural discourse—especially some of the most well-known statements of design philosophy—are essentially untestable. For example, the Renaissance–baroque view that architecture should adhere to the principles of beauty and harmony cannot be tested.

[2]It is interesting to note that at least one architectural educator has characterized the nature of scientific and architectural theories in the reverse way. Dunham-Jones (1988) has argued that scientific theories are prescriptive in that they "become rules" and dictate methods; in contrast, she maintains that architectural theories have traditionally been focused on "the construction of new hypotheses." Given this reversal of definitions, it is tempting to speculate that it is a lack of familiarity with the discipline in question (whether it be architecture or behavioral science) that leads to the conclusion that its theory is "prescriptive."

Nor is it possible to test Le Corbusier's assertion of modernist dogma in his statement that the exterior should be "the result of the interior" (1960, p. 11). However, to say that these manifestos and polemical proposals for built form constitute the full measure of architectural theory is to confuse the prescriptive theory of the profession with the body of theory that pertains to the discipline of architecture. And in a similar vein, a mistaken view of environment–behavior research would result from a failure to distinguish between the prescriptive theory that may be implicit in design guides or postoccupancy evaluations (i.e., the insistence that behavior principles are fundamental to these activities) and the explanatory theory upon which many environment–behavior studies are based.

Taking the previous examples a bit farther then, we can say that while it is not feasible to test a belief in beauty, it is possible to test whether the use of Renaissance–baroque principles of hierarchical ordering actually produce buildings that are interpreted as beautiful by a given set of people. Similarly, while it may not be possible to test the value of functional expressionism, it is, in fact, possible to test whether building composed in that way are actually interpreted in terms of either their functional components or their function as a whole. Put another way, many of the design principles described in architectural discourse constitute implicit hypotheses.

Based on this analysis of architectural theory, it can be argued that, despite the fact that most architectural theorists have not been particularly inclined to test theory, many architectural "theories" are, indeed, "testable in principle." And if this is the case, then the domains of architectural and environment–behavior theory need not be conceptualized as unrelated domains.

Finally, the preceding historical overview of architectural discourse in relation to evolution of the scientific epistemology suggests that the lack of an apparent relationship between architectural and environment–behavior "theory" may be, in part, a function of temporary philosophical disjuncture. To put this point in the most oversimplified terms, design research theory can be characterized as operating primarily within a modern mode of thought, whereas architectural theory of the last 10 years (if not longer) has operated primarily within the postmodernist perspective.

To advance this argument, it is necessary, however, to consider the nature of the philosophical assumptions that are necessarily implicit in any theory. Specifically, Moore (1988) has argued that all theories presuppose at least three sets of assumptions: (1) ontological, which concern the nature of being and reality; (2) epistemological, which pertain to the nature of knowledge; and (3) methodological, which have to do with how one investigates phenomena. Taken together, these assumptions provide the philosophical basis for any given theory.

In this light, it is possible to characterize the underlying philosophical assumptions of environment–behavior theory, at least in the North American literature, as consistent with the premises of modern thought. For example, Ittelson (1989) has described the environment–behavior research field as being based on a "materialistic and mechanistic" set of assumptions. And similarly, Stokols (1977) described the field as offering the potential to develop as "a new, theoretically coherent discipline bridging *the behavioral and design sciences*" (p. 3, our emphasis). And in a more critical vein, Gergen has suggested that psychology has failed to come to terms with the psychological implications of art because most research in this area has implicitly accepted the empiricist (or neo-enlightenment) assumptions of modern thought. Gergen's argument is that psychology must be wrenched from its old habits and draw from the "'post-modern dialogue'" (1988, p. 2).

In contrast, as the analyses in the previous section have indicated, contemporary architectural theory generally accepts the premises of postmodern thought. It thereby stands intentionally as a challenge to the modern mode of thought, not only with respect to its implications for physical form but its ontological–epistemological framework as well. Although it is certainly true that some theorists (e.g., Colquhoun, 1988; Dovey, 1988; Lawrence, 1989; Rochberg-Halton, 1986) have questioned the extent to which either structuralism or phenomenology (as currently constituted) can serve as an effective foundation for breaking fundamentally with the philosophical premises of modernity, it is nevertheless true that recent architectural theory has at least attempted to do so. Indeed, much of the discourse in contemporary architectural theory has questioned the extent to which science (as it is typically defined in modern thought) can usefully describe or explain the essence of our relationship to built form. Examples of this tendency include Macrae-Gibson's concept (1985) of "poetic logic" (which echoes the pre-enlightenment notions of Vico) and Perez-Gomez's discussion (1983) of the transcendental dimension of architecture.

In summary, then, we would suggest that the apparent discontinuity between the domain of architectural theory and that of environment–behavior research is illusory. In part, this illusion derives from a failure to recognize that the distinction between the *profession* and the *discipline* applies equally well to both architecture and the environment-behavior field. As a consequence, the fact that both fields make use of both explanatory and normative "theory" has been obscured. Even more important, the illusion of discontinuity between the two domains of theory has been exacerbated by what may ultimately be a temporary philosophical disjuncture, namely, the disjuncture between the modernist assumptions of environment–behavior theory and the postmodern inclination of architectural theory.

TOWARD A RECONSIDERATION OF THE DOMAIN OF THEORY IN ENVIRONMENT–BEHAVIOR RESEARCH

THE MAJOR THEMES OF ARCHITECTURAL THEORY

Since the main thesis of this chapter is that environmental design research must—if it is to be seen as relevant to design professions—incorporate architectural theory, the next logical step is to look at the content of these literatures.[3] Thus, this section (1) presents an overview of contemporary architectural discourse, with particular reference to three classificatory criteria—physical attributes, levels of analysis, and temporal focus—and (2) analyzes examples of environment–behavior studies that relate, either implicitly or explicitly, to architectural theory.

The identification of key physical attributes. Within contemporary architectural theory, it is possible to identify a variety of ways that the built environment is differentiated for analytical purposes. Since a central argument to this chapter is that environment–behavior researchers have not been altogether successful in identifying the key physical attributes of built form, such a classification can provide valuable insights about how physical attributes of built form may be studied. Five major formal properties of the built environment found in architectural theory are reviewed: style, composition, type, morphology, and place.

The issue of *style*, defined as "a conscious system of design, a visual code based on tectonic preference, a post-vernacular language of form," has remained central to architectural theory since the Renaissance (Crook, 1987, p. 13). In *The Dilemma of Style*, Crook (1987) argues that, from neoclassicism to postmodernism, the nature and the necessity of style has remained a major concern for generations of architects and theorists.

---

[3]Although an extensive review of architectural theory throughout the four periods identified in the previous chapter section (Renaissance-baroque, premodern, modern, and postmodern) might usefully serve to define the themes that have remained central to this body of literature, such a discussion is unfortunately beyond the scope of this chapter. As a consequence, only contemporary architectural theory will be referenced in the remainder of this chapter.

Architectural *composition*—the way in which parts or elements of an architectural whole are combined—typically includes issues of geometry, order, hierarchy, and proportion in the assembly of different components or elements of architecture. The issue of composition has, like style, remained central in the history of architectural ideas, and contemporary architectural theorists continue to give it an important place in their discussions. Tzonis and Lefaivre's *Classical Architecture* (1986) and Venturi's *Complexity and Contradiction in Architecture* (1966) are examples of this enduring consideration of compositional issues. Tzonis and Lefaivre discuss the way classical buildings were put together as formal structures by means of their generative rules or "poetics of composition"; Venturi explores how architectural composition can be used to generate specific interpretations of architectural form.

A third basis for analyzing formal properties is the concept of type, a notion which has been central to architectural discourse since the Enlightenment. *Type* can be defined as a group of objects characterized by the same formal structure. Extensively used in premodern architectural theory, the notion of type was popularized in its postmodern renewal by Aldo Rossi's *Architecture of the City* (1966; English edition, 1982). Since then, it has been the subject of analyses by numerous architectural critics and historians (e.g., Colquhoun, 1981; Moneo, 1978; Vidler, 1977, 1987). Recent writing such as Geist's *Arcades: The History of a Building Type* (1983) and Krier's *Elements of Architecture* (1983) acknowledge the importance of this notion in architectural literature.

A fourth construct central to architectural theory is the concept of *morphology*, the origins of which can be found in urban theory. The concept of urban morphology, which is frequently linked to analyses of typology, emphasizes those aspects of architecture which can be analyzed as *a system of spatial relations*. This notion, which builds substantially upon the work of numerous Italian architectural theorists, was also popularized by Rossi's *Architecture of the City* (1982). Krier's *Urban Space* (1979) and Rowe and Koetter's *Collage City* (1978) represent other examples of this kind of analysis.

Finally, the concept of place has emerged in architectural theory in the last 15–20 years. Norberg-Shultz (1980a) defines *place* as a "totality made up of concrete things having material substance, shape, texture, and color . . ." (p. 6) and which together determine the environmental character, the essence of a place, its atmosphere. For Norberg-Shultz, place is a total phenomenon that cannot be reduced to either physical form or spatial relationships. Moreover, the growing popularity of the concept of place is evidenced by numerous other works, such as Charles Moore's *Place of Houses* (1979) and the journal *Places*.

The temporal dimension of analysis.   A second criterion by which architectural theory can be classified has to do with the temporal—diachronic or synchronic—dimension of analysis. A *diachronic* analysis can
be understood as representing a longitudinal perspective, while a *synchronic* analysis can be conceived as a cross-sectional analysis. For instance, Hersey's *The Lost Meaning of Classical Architecture* (1988) is a diachronic analysis of the classical language of architecture, while Tzonis
and Lefaivre's *Classical Architecture* (1986) is a synchronic (atemporal)
analysis of the essential compositional rules of the classical language of
form.

Levels of analysis.   A third basis of classifying the body of literature
in architectural theory is represented by the level of analysis adopted.
Two major perspectives can be found: the first focuses on the strictly
formal or structural properties of the built form, and the other focuses
on the symbolic content or meaning associated with specific formal arrangements. The latter focus is found in the work of Broadbent (1977)
and Groat (1981, 1983, 1987a), both of whom insist on the significance of
the semantic aspects of architecture—its capacity to carry meaning—in
architectural theory during the last 20 years. More specifically, Broadbent *et al.*'s book, *Signs, Symbols, and Architecture* (1980) consists almost
entirely of semiological analyses from architectural critiques that consider how the formal or structural properties of the built form carry meaning. And in a similar way, Bonta's *Architecture and Its Interpretation* (1979)
analyzes modern architecture from the point of view of its changing
meaning.

The three general classificatory criteria defined in this section—
physical attributes, the temporal dimensions, and, levels of analysis—
are complementary rather than mutually exclusive. Thus, a given research study can be classified according to each of the three criteria. For
instance, it can focus on the *meaning* (level of analysis) of a specific *style*
(physical attribute) in *synchronic* (temporal) terms.

Very few environment–behavior researchers have explicitly built
upon architectural theory to define their research designs. In particular,
only a few environment–behavior researchers have explicitly used any
of the five categories of physical attributes identified in the architectural
literature. With respect to levels of analysis, environment–behavior researchers have tended to focus on the content–meaning aspects of architectural production, often to the exclusion of its formal properties. Finally, researchers have typically focused on synchronic (or cross-sectional)
analyses. In contrast, theoretical discourses in the architectural literature have tended to focus strictly on the formal or structural aspects of
architecture, neglecting or ignoring people's relationships to the built

environment. Nevertheless, we argue that there is both a need for and an opportunity to reconnect the abstract formal analyses of architectural theory with an understanding of the way people experience these forms as cultural artifacts. This lack of commonality in research focus (particularly with respect to levels of analysis) may be, in part, one of the primary symptoms of the lack of communication between the domain of architectural theory and environment–behavior research.

The environment–behavior research studies reviewed in the next sections are unusual in that they have investigated people's relationships to the built environment from a perspective that incorporates—though sometimes implicitly rather than explicitly—the architectural literature, thereby linking architectural theory with environment–behavior research theory. In other words, these studies have adopted conceptual frameworks that are compatible with existing architectural theory by including one or more of the themes central to architectural discourse. The various studies, reviewed below, are classified according to the five categories of physical attributes identified earlier: style, composition, type, morphology, and place.

STYLE

Environment–behavior research studies on the issue of style have tended to focus on either (1) people's *recognition or formal categorization* of defined architectural styles, (2) *people's stylistic preferences for a given building type,* or (3) the *social meaning* associated with housing of particular architectural styles. Examples of the first group include studies by Perianez (1984), Groat (1982), Espe (1981), and Verderber and Moore (1977). Perianez (1984) tested the existence of a system of stylistic classification among people who were asked to judge a range of architectural imageries. The results indicated that people can easily classify "stylistic periods" and propose logical links between them.

Similarly, Groat (1982) investigated architects' and nonarchitects' abilities to distinguish between modern and postmodern buildings by asking them to sort photographs of modern, transitional, and postmodern buildings. The results of her analysis indicated that, while the modern–postmodern distinction played a significant role in the construct systems of half the architects, this stylistic distinction was evident in the responses of only one of the nonarchitects. Moreover, architects and nonarchitects tended to make use of different classificatory criteria.

Other studies, e.g., Espe (1981) and Verderber and Moore (1977), have also examined the extent to which expert categorizations of formal stylistic qualities are perceived by laypeople. To be specific, Espe's study examined differences in the perception of national socialist and classicist architecture. The results indicated that the respondents not only discriminated between the Nazi and classical architectural styles but also consistently associated certain formal attributes with specific meanings. Similarly, Verderber and Moore (1977) tested the hypothesis that "high style" and "popular" architecture consists of two distinct stylistic expressions. Their analyses demonstrated that all three respondent groups (representing different socioeconomic and educational backgrounds) "did cognitively view high architecture as different from popular architecture" (p. 337).

A second area of interest identified by empirical studies on style concerns people's stylistic preferences within a given building type. For instance, in a study of city hall imagery by Lee (1982), people were asked to indicate their preferences for a city hall among a set of photographs. Older buildings were found to represent an appropriate image for a city hall, and the issue of style emerged as one of the major concerns in the participants' responses to the photographs. Similarly, a study by Schermer (1987) explored people's preferences—in terms of image, style, and symbolism—among suggested alternatives for a school of architecture and for a small town library. The results suggested that people have strong and clear attitudes about what stylistic expressions are appropriate to the proposed facilities.

A third area of interest in environment–behavior studies on style concerns the social meanings associated with specific building styles. For instance, a study by Nasar and Kang (1989) investigated people's association of different architectural styles in small office buildings with specific evaluative and functional meanings. People were found to consistently associate particular office styles with specific meanings. Similarly, a study by Cherulnik and Wilderman (1986) investigated people's current perceptions of nineteenth-century Boston houses, specifically in terms of the residents' social status. The results indicated that respondents' answers were consistent with the socioeconomic category of original owners; this suggests that social meanings are, indeed, associated with specific formal features that may serve as symbols of residents' status.

One problematic aspect in many of the empirical studies of style is the lack of specificity with which the various styles are operationalized. One example of a successful use of architectural analyses within an empirical study of style was provided by Després (1987). This study analyzed the meaning of the vernacular French Canadian rural house as reproduced in Quebec's contemporary suburbs. Although this study

FIGURE 10. Comparative stylistic analysis of a vernacular housing type (top) and its reproduction in contemporary Quebec suburbs (bottom). (Després, 1987)

was primarily concerned with the symbolic meaning of a specific style, it
nevertheless dealt directly with the physical properties of style.
Through a structured content analysis of the elevations of vernacular
and neovernacular houses, the author established discriptors for both
the original style and its "neo" counterpart. The two stylistic expressions were then compared in their compositional, functional, and

symbolic aspects in order to establish the extent to which the neostyle had remained "true" to the original style. The results indicated that the popularity of the neovernacular style could be explained neither by the desire for stylistic authenticity nor by nostalgia for a lost mode of dwelling. Supported by data on the sociopolitical context in Quebec in the 1960s and 1970s—the period during which the neovernacular house was introduced in Quebec's suburbs—the author concluded that the neovernacular style's popularity was primarily a consequence of the semantic associations of the neo-quebecois style with the French-Canadian culture.

Després' study is also relatively unique among environment–behavior studies of style in that it incorporates a diachronic dimension of analysis. In contrast, as most of the examples of research reviewed in this section indicate, environment–behavior research on style has typically taken a primarily synchronic temporal focus. Although a number of the studies have, in fact, concerned themselves with perceptions of historical styles, the essential objective of the research is an analysis of contemporary interpretations. Among these examples, then, Després' work is significant in that it draws upon sociocultural literature to incorporate an historical analysis of the political trends that seem to have motivated both a stylistic appropriation and a consensual interpretation of form.

COMPOSITION

Examples of empirical studies dealing with compositional issues are relatively rare in the environment–behavior literature. Most of the research studies can be classified into one of two major thematic foci. The first of these includes all of those studies that have been based on a particular aspect of composition: the concept of complexity. The second, and much smaller, group consists of those studies that have attempted to explore the relative significance of certain physical dimensions of the environment. In addition, the issue of composition has been found to be a powerful variable in research on other significant topics in environment–behavior studies (e.g., Groat, 1987a; Purcell, 1984a).

With regard to research on the role of complexity in architecture, most studies derive, at least implicitly, from the work of Berlyne (1971). As is well known, his model of "aesthetic appreciation" combines concepts from both psychophysiology and aesthetic theory. Within the context of this theoretical model, Berlyne argued that complexity represented one of the most significant variables, if not *the* most significant, in any art form. According to Berlyne, a "pattern is considered more complex, the larger the number of independently selected elements it contains" (1971, p. 149). Although Berlyne's model has been subjected to substantial criticism (e.g., Crozier & Chapman, 1984; Wohwill, 1976), it has nevertheless formed the basis of several analyses of contemporary architecture in the environment–behavior literature (e.g., Prak, 1977; Rapoport & Kantor, 1967). It has also served as the basis for a variety of empirical studies on the role of complexity in urban aesthetics (e.g., Geller et al., 1982; Nasar, 1983).

Within the context of this discussion, the most crucial issue raised by this body of work concerns the operationalization of the term complexity. For example, Nasar (1983) equates the factor "diversity" (derived from ratings of design professionals) with complexity, and Geller et al. (1982) provide a post hoc definition of complexity based on their subjects' responses to open-ended questions. Given these ambiguities, it is difficult to assess the comparability of these studies to Berlyne's model, to each other, and especially to the concept of complexity in the design literature. This lack of comparability also lends support to Wohlwill's (1976) argument that complexity ought not to be considered a unidimensional attribute.

In a different vein, Purcell's research (1984a) on the role of categorization in cognitive processes demonstrates the potential significance of complexity as an important compositional factor in aesthetic experience. In analyzing the lay subjects' responses to various examples of housing, Purcell concludes that "stimuli which are both complex and perceptually well-organized are intrinsically interesting and attractive" (1984a, p. 209). Although Purcell, like the previous examples, does not present a precise definition of complexity, his conclusions implicitly relate these concepts to analyses of compositional principles in the architectural literature.

The second thematic category of research consists of those studies that have explored how the size and proportion of certain forms have influenced the perception of interior spaces. For example, Sadalla and Oxley (1984) explored the influence of the ratio length/width of a room on its perceived spaciousness. The authors concluded that the rectangular rooms (defined by greater length/width ratios) were perceived as more spacious than rooms which more closely approximated square proportions. A second example of this kind is represented by Canter and Wools's (1970) study of the effect of roof angle, window size, and type of furniture on room friendliness. Their findings suggested that roof angle accounted for over 50 percent of the variance and furniture type accounted for 35.

A major limitation of these studies, however, is that the specific physical features are conceived as independent elements of physical form. While this has obvious advantages for simplifying research design, it ignores some basic realities of design—namely, that in designed environments, such features as the asymmetry of fenestration or the angle of the ceiling section are not designed independently of each other, but as parts of a system. It is, then, the compositional system which requires description and analysis.

One example of research that has incorporated an analysis of compositional *systems* is Groat's (1984, 1987b) research on people's perception of contextual compatibility. She asked 97 respondents (73 non-architects and 24 design review commissioners) to sort a set of 25 infill projects simulated through color photographs. Her initial results indicated that the physical features that contributed most significantly to the perception of compatibility had to do with facade design, as opposed to either site organization or massing; and secondly, that the most preferred contextual relationships were those that embodied a relatively high degree of replication. However, even though the three postmodern buildings in the study showed concern for the articulation of the facade

FIGURE 11. The Allen Art Museum at Oberlin College, Oberlin, Ohio (1973–1977), by Venturi Rauch and Scott Brown is an example of a postmodern design strategy to the problem of contextual compatibility. Courtesy of Venturi Rauch and Scott Brown.

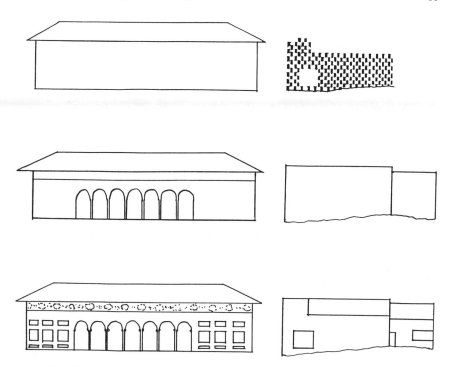

FIGURE 12. This analysis of the facades of the original museum and its addition demonstrates that the basic principle of hierarchical ordering typical of premodernist buildings has been inverted in the postmodern addition. (Groat, 1987b)

and its details, they were usually not perceived as being particularly compatible with their context. These anomalies in the results suggest that replication and/or reinterpretation of specific design features, while important, may not be the only critical design principle in the perception of contextual compatibility. To account for this variation, Groat reanalyzed the facade composition of the buildings and discovered that the principle of hierarchical ordering (which is typical of premodernist facade composition) played an important role in the perception of contextual compatibility.

TYPE

Type is a concept which represents a potentially significant point of intersection between architectural theory and environment–behavior research. Indeed, several researchers (e.g., Groat, 1987c; Vernez-

Moudon, 1987; Robinson, in press) have argued strongly for the usefulness of type as an analytical category in environment–behavior research. In addition, a number of empirical studies have explored the relationship between architectural types and people's experience of them—although with different degrees of awareness of the central role of this notion in architectural theory.

To date, however, the concept of type has been defined in fundamentally different terms in the two sets of literature. Generally speaking, the literature on type in architectural theory has focused almost exclusively on *formal* types (i.e., L-shaped vs. courtyard buildings), whereas the environment–behavior literature has tended to focus on *functional* types (i.e., hospitals vs. prisons). Although some authors have defined type in terms of the confluence of the two perspectives (e.g., Vernez-Moudon, 1987), few of the analyses in either realm of discourse have achieved this degree of resolution.

Within the environment–behavior literature, the various studies can be classified into three major sets: (1) analyses of built form, from a primarily diachronic perspective, which infer meanings from the structural properties of the forms themselves; (2) studies concerning people's cognitive representation of different building types; and (3) studies exploring the links between building types and specific personal or cultural meanings.

Research in the first category of typological studies—i.e., those which infer cultural meanings from specific formal arrangements—includes the work of Lawrence (1986) and Goodsell (1986). More specifically, Lawrence (1986) investigated the spatial organization of popular housing in Switzerland between 1860 and 1960. Using archival material, he analyzed floor plans of buildings in terms of the transformations of private, communal, and public space associated with each dwelling. Lawrence was not only able to identify a typology of dwelling but also to demonstrate how cultural interpretations of the public and private realms are embedded in built form. In a similar way, Goodsell (1986) analyzed both plans and photographs of city council chambers in 75 city halls in the United States and Canada. Among the specific physical features he catalogued were the overall composition of space, the presence and types of heavy furniture or anchored floor barriers, the character of decorative features, and the types of portable objects. His analysis identified three eras or periods in the expression of civic authority— 1865–1920, 1920–1960, 1960–present; each was represented by a distinct set of physical attributes.

A second set of typological studies in the environment–behavior literature focuses on the exploration of people's cognitive representations of specific building types. These studies have tested the existence of a typology of buildings, based on functional use. Purcell, in particular, has done extensive work on this issue (1984a, 1984b, 1984c). His research program explores the role that prototypes (i.e., categorical exemplars) play in the process of aesthetic appreciation of environmental form. He hypothesized that people would identify a prototype(s)—or ideal type(s)—among a series of buildings belonging to the same category. In testing his hypothesis within the context of single-family houses and churches in two separate research projects, he asked people to rate photographs for goodness-of-example and for interest. In each case, Purcell was able to identify the prototype—or ideal type—for a given building function.

Similarly, although less firmly based on a model of cognitive processes, Young's study (1979) attempts to identify prototypical features of a much broader spectrum of building functions. A sample of architectural and engineering students were asked to sort sketches of buildings into six building function categories: house, flat, church, office building, factory, and school. These data enabled Young to identify a set of morphic features typically associated with each building type.

In a slightly different vein, Krampen (1979) sought to determine the extent of formal differentiation required for the identification of building function. Respondents were asked to sort four sets of sketches (each set representing an increasing level of visual articulation in the buildings) into six functional types: office building, factory, tenement, church, school, and individual house. The analysis showed that the respondents did not succeed in distinguishing the six categories of buildings for the first two sets of drawings with low levels of detailing. However, when windows were added, participants were able to categorize the buildings, with the exception of the schools. Only when given the fourth set of drawings were they able to complete the task. These findings suggest that a threshold of architectural detail is necessary to identify the social function of a building.

The third group of typological studies explores the links between building types and specific personal or cultural meanings. For example, Robinson (1988) has explored the link between the perceived quality of various types of housing and their physical attributes (as measured in terms of homelike and institutionlike features). Her research strategy consisted of three overlapping techniques: (1) a checklist of 236 individual features of homelike and institutionlike residential settings; (2) a semantic differential evaluation, based on slides from the various settings; and (3) a sorting task of black-and-white photographs of the residential settings. Her results indicated that residential settings, both in

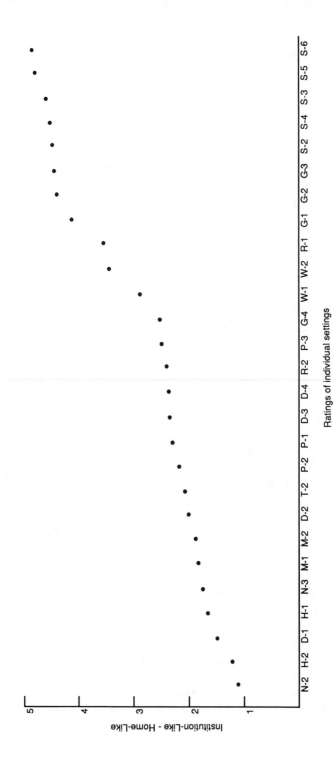

FIGURE 13. Mean institutionlike—homelike ratings of slides of building exteriors. (Derived from research by Julia Robinson, 1988) D:dormitory; G: group home; H: hospital; M: mid-highrise; N: nursing home; P: public housing; R: rooming house; S: single family house; T: townhouse; W: walk-up apartment.

terms of physical features and perceived quality, can be characterized as falling along a continuum of homelike–institutionlike.

Other research studies on typology have investigated a variety of other building types. For example, Kantrowitz, Farbstein, and Schermer (1986) investigated the meaning of post office among customers and employees. Slides of post office building exteriors, lobbies, and interior work areas were presented to the participants, who were then asked to record three words that described the important psychological aspects and physical features of a post office.

Morphology

Environment–behavior research studies that consider the morphological properties of the built environment are relatively rare. This section presents two studies which have attempted to go beyond the descriptive analysis of morphology (more typical of architectural theory) and to incorporate an understanding of how those spatial relationships are experienced and interpreted within a sociohistorical context.

Perhaps the best and most representative example of this approach is Vernez-Moudon's analysis of a nineteenth-century inner-city neighborhood in San Francisco (1986). Her research incorporates both synchronic and diachronic analyses of the house as a "unit of building, as a cell of city organism." This was accomplished through a comparative analysis of maps showing the land subdivision, the footprints of buildings, the height of buildings, the principal land uses, and the residential use for each property. Fifty-six blocks and three different times in the history of the neighborhood (1899, 1931, 1976) were selected for analysis. The several stages of the analysis consisted of: "(1) identifying different generations of buildings, (2) identifying types of buildings within each generation, and (3) identifying principles of design that governed the form and layout of common buildings" (p. 245). Although the study does not deal directly with people's experience and interpretation of their environment, Vernez-Moudon nevertheless argues that the human aspects of architecture are embedded in the built form and in its evolution and transformation in time:

> Our analyses did not and could not lead to the belief that "building comes before people." We did not present our typologies of architectural and neighborhood space as definite or static entities, but, on the contrary, as constantly evolving in the hands of their owners and residents. Because we looked at change in the physical elements of neighborhoods, we saw those elements as the tools as well as the representation of culture. (Vernez-Moudon, 1986, p. xii)

A second, and quite different, example of morphological analysis is Hillier and Hanson's book, *The Social Logic of Space* (1984). Their analysis is primarily theoretical in focus and constitutes an attempt to account for variations in spatial arrangements or morphological types. They identify the most fundamental property of the built environment as the "ordering of space into relational systems embodying social purposes" (p. 2), the most significant of which are those having to do with the economic and political structure of society. Although Hillier and Hanson's analyses provide a potentially useful theoretical basis for more specific research on people's experience of the morphological aspects of the built environment, this potential has yet to be realized in the environment–behavior research literature.

PLACE

The concept of place represents an important and especially significant confluence of the environment–behavior literature and architectural theory. More specifically, the concept of place is distinct from the four categories of physical attributes described in the previous segments of this chapter in that it so clearly embodies a combination of both physical and semantic qualities. In fact, this merging of the two qualities is evident in both sets of literature.

Recently, Sime (1986) has attempted to summarize the broad range of relevant literature on the concept of place. Within the environment–behavior research literature, he has identified two major perspectives. The first of these is represented by the body of work in human geography (e.g., Relph, 1976; Seamon, 1982; Tuan, 1977), which derives chiefly from the phenomenological orientation of the German philosopher Heidegger. In this respect, the human geographers share a common intellectual source with a number of the architectural theorists (e.g., Norberg-Shultz, 1980a) who have also articulated the concept of place.

The second perspective in the environment-behavior field is the empirically based perspective of environmental psychology. This is perhaps best exemplified by the work of David Canter (1977, 1983). While the proponents of this perspective believe it is important to measure empirically the relationship between people and their physical surroundings (Sixsmith, 1983), the phenomenologists argue that this approach eliminates too much of the emotional content linking people to a place.

Despite these differences in the epistemological basis of place, it is nevertheless remarkable that both Canter and Relph have defined place in terms of the same three, interlocking components: physical attributes, activities, and conceptualizations or meanings (Canter, 1977, p. 158; Relph, 1976, p. 47). In comparing the two approaches to place, Sime (1986) has pointed out that while the phenomenological perspective can legitimately by criticized for taking an overly subjective view of the evaluation of places, the environmental psychology perspective can be criticized for paying inadequate attention to "the objective physical environment which architects have to manipulate" (p. 55).

Although only a few environment-behavior studies have been explicitly informed by either the Canter or Relph model of place (e.g., Donald, 1985), an example of a research study that is directly informed by Canter's model of place is provided by Sixsmith (1986). Respondents in this study were asked to sort their own descriptions of all past, present, and possible ideal homes. Sixsmith's analysis revealed 20 categories of home, grouped under the labels of personal home, social home, and physical home. She concluded that home must be conceived as a multidimensional phenomenon, and that "each home features a unique and dynamic combination of personal, social, and physical properties and meanings" (p. 294).

One problematic aspect of the place perspective in environment-

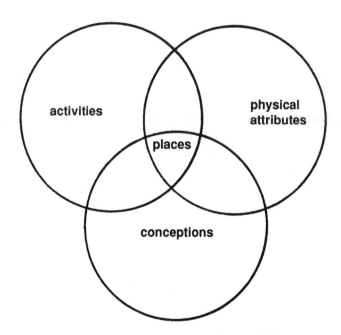

FIGURE 14. The model of place. (Canter, 1977)

behavior studies is the lack of attention given to physical form in the definition of the "sense of place." In this light, Després' investigation (in press) of the meaning of home for nontraditional households represents an important exception. Després argues that the meaning and experience of places involves an interplay among the built form itself, individual motivations, and societal forces. For this reason, the morphological aspects of shared dwellings are analyzed—independently of the occupants' evaluations of their housing units—for their potential contribution in enhancing or limiting the experience of being-at-home.

IMPLICATIONS FOR FUTURE RESEARCH AND APPLICATIONS

This chapter has presented three interrelated arguments. First, it has been argued that environment–behavior research, by definition, must concern itself with the description and explanation of physical form. Second, it has also been suggested that the body of literature usually termed "architectural theory" can provide an appropriate basis for studying the significant physical attributes of the built environment. And finally, the preceding chapter section has demonstrated the potential usefulness of this approach by identifying a number of relevant examples of specific environment–behavior studies.

These examples were initially introduced in terms of three classificatory criteria: the specification of key physical attributes, the temporal dimension of analysis, and the level of analysis. These criteria serve as a useful conceptual framework for evaluating both the present and future of environment–behavior research. With respect to the first of these criteria, five modes of analyzing physical form were identified: style, composition, type, morphology, and place. As the preceding literature review has revealed, it is clear that environment–behavior researchers have tended to focus almost exclusively on the analysis of style and type, thus leaving the concepts of composition, place, and especially morphology relatively unexamined. Moreover, among the studies on style and type, few have actually been conceptualized with adequate reference to the rich and fine-grained definition of these concepts found in architectural theory. The obvious conclusion here is that environment–behavior research would be usefully informed by (1) more research on the topics of composition, place, and morphology and (2) the elaboration of research on style and type through the use of concepts from architectural theory.

With respect to the second criterion, the temporal dimension of analysis, it is clear that the vast majority of existing research is essentially synchronic. Although synchronically focused research can, of course, be valuable in its own terms, we argue that the explanatory power of such research could be substantially augmented by the development of diachronically focused research. Indeed, this argument echoes those of other authors, both in architectural theory and environment–behavior research (e.g., Colquhoun, 1981; Vernez-Moudon, 1987). As earlier chapter sections have indicated, a few researchers have already begun to explore the potential of diachronic analyses; it is time for others to take up this challenge as well.

Although the third classificatory criterion, level of analysis, was initially described primarily with reference to the literature on architecture theory, it is equally useful as a basis for evaluating environment–behavior research. Indeed, the situation in environment–behavior research represents the converse of the situation in architectural theory. Specifically, whereas architectural theorists have tended to concentrate on the formal properties of built form to the exclusion of its meaning, environment–behavior researchers have tended to focus on the meaning of built form rather than its formal properties. Obviously, both enterprises would be usefully served by incorporating both levels—formal properties and meaning—in their analyses.

Certainly, although the specific examples of environment–behavior research that have been reviewed here are relatively few in number, it is significant that most of these studies have been conducted within the last five years. Such a phenomenon would suggest that an increasing number of researchers see the value of drawing from the literature on architectural theory. Indeed, this is likely to be especially true for environment–behavior researchers who work primarily with architects in either academic or consulting situations.

Over the years, environment–behavior research has been repeatedly criticized for its lack of applicability to the design professions in general and to architecture in particular. In response to these concerns, researchers have developed a number of strategies—including action research, design guidelines, as well as programming and evaluation studies—for making more effective links to architecture and the other design fields. However, it is our belief that these efforts reflect a rather limited view of the potential relationship between environment–behavior research and architecture. Instead, we argue that environment–behavior research must concern itself not simply with applications to the profession of architecture but to the discipline as well (Anderson, 1988). It is in this context, in particular, that we believe the literature on architectural theory may serve as a potentially invaluable source for the development of innovative approaches to environment–behavior research.

Finally, and most important, the potential of this approach should be viewed as a complement to—rather than in opposition to—the ongoing traditions of environment–behavior research.

REFERENCES

Alberti, L. B. (1988). *On the art of building in ten books*. Cambridge, MA: MIT Press.
Anderson, S. (1988, November). *Some themes for a symposium on Ph.D. education in architecture*. Paper presented at the Symposium on Doctoral Education in Architecture, University of Michigan, Ann Arbor.
Berlyne, D. (1971). *Aesthetics and psychobiology*. New York: Appleton-Century-Crofts.
Bonta, J. P. (1979). *Architecture and its interpretation: A study of expression systems in architecture*. London: Lund Humphries.
Broadbent, G. (1977). A plain man's guide to the theory of signs in architecture. *Architectural Design, 7–8*, 474–482.
Broadbent, G., Bunt, R., & Jencks, C. (1980). *Signs, symbols and architecture*. New York: Wiley.
Brolin, B. (1976). *The failure of modern architecture*. New York: Van Nostrand.
Canter, D. (1977). *The psychology of place*. London: Architectural press.
Canter, D. (1983). The purposive evaluation of places: A facet approach. *Environment and Behavior, 15*, 659–698.
Canter, D., & Wools, R. (1970). A technique for the subjective appraisal of buildings. *Building Science, 5*, 187–198.
Cherulnik, P. D., & Wilderman, S. K. (1986). Symbols of status in urban neighborhoods. *Environment and Behavior, 18*, 604–622.
Colquhoun, A. (1981). *Essays in architectural criticism*. Cambridge, MA: MIT Press.
Colquhoun, A. (1988, February). Postmodernism and structuralism: A retrospective glance. *Assemblage, 5*, 7–15.
Crook, J. M. (1987). *The dilemma of style*. Chicago: University of Chicago Press.
Crozier, W., & Chapman, A. (Eds.). (1984). *Cognitive processes in the perception of art*. New York: North-Holland.
Davis, D. (1978, November 6). Designs for Living. *Newsweek*, pp. 82–91.
Després, C. (1987). Symbolic representations of the suburban house: The case of the neo-Québecois house. *Public environments*. Washington, DC: Environmental Design Research Association.
Després, C. (in press). *The meaning and experience of home in the context of non-family households living in shared housing*. Ph.D. dissertation, Department of Architecture, University of Wisconsin, Milwaukee.
Donald, I. (1985). The cynlindrex of place evaluation. In D. Canter (Ed.), *Facet theory: Approaches to social research* (pp. 173–204). New York: Springer-Verlag.
Dovey, K. (1988). Place, ideology, and postmodernism. In H. van Hoogdalem, N. Prak, T. van der Voordt, & H. van Wegen (Eds.), *Looking back to the future* (pp. 275–285). Delft, the Netherlands: Delft University Press.
Dunham-Jones, E. (1988). *Architectural theory: Prescription or proposal*. Paper presented at the annual meeting of the Association of Collegiate Schools of Architecture, Miami, Florida.
Eco, U. (1984). *Postscript to* The name of the rose. New York: Harcourt Brace Jovanovich.

Espe, H. (1981). Differences in the perception of national socialist and classicist architecture. *Journal of Environment Psychology, 1,* 33–42.

Foster, H. (1984). (Post) modern polemics. *Perspecta: The Yale Architectural Journal, 21,* 144–153.

Geist, J. F. (1983). *Arcades: The history of a building type.* Cambridge, MA: MIT Press.

Geller, D. M., Cook, J. B., O'Connor, M. A., & Low, S. K. (1982). Perceptions of urban scenes by small town and urban residents. In D. Bart, A. Chen, & G. Francescato (Eds.), *Knowledge for design* (pp. 128–141). Washington, DC: Environmental Design Research Association.

Gergen, K. (1988, August). *Understanding as literary achievement.* Presidential address to division 10, psychology and the arts, annual meeting of the American Psychological Association, Atlanta, GA.

Giovannini, J. (1988, June 12). Breaking all the rules. *New York Times Magazine,* pp. 40–43, 126, 130.

Goodsell, C. T. (1986). The social meaning of civic space. In J. Wineman, R. Barnes, & C. Zimring (Eds.), *The cost of not knowing.* Washington, DC: Environmental Design Research Association.

Groat, L. (1981). Meaning in architecture: New directions and sources. *Journal of Environmental Psychology, 1,* 73–85.

Groat, L. (1982). Meaning in post-modern architecture: An examination using the multiple sorting task. *Journal of Environmental Psychology, 2,* 3–22.

Groat, L. (1983). The past and future of research on meaning in architecture. In D. Amedeo, T. B. Griffin, & J. J. Potter (Eds.), *Proceedings of the fourteenth international conference of the Environmental Design Research Association. Lincoln, NB* (pp. 29–35). Washington, DC: Environmental Design Research Association.

Groat, L. (1984). Public opinions of contextual fit. *Architecture: The AIA Journal, 73* (11), 72–75.

Groat, L. (1987a). Recent developments in architectural theory: Implications for empirical research. *Journal of Environmental Psychology, 7,* 65–76.

Groat, L. (1987b). Contextual compatibility: An issue of composition, not replication. In T. Beeby & A. Plattus (Eds.), *Architecture and urbanism* (pp. 317–322). Washington, DC: Association of Collegiate Schools of Architecture.

Groat, L. (1987c). Typology: A basis for enhancing the domain of architectural research. In R. Shibley, W. Mitchell, W. Pulgram, & R. McCommons (Eds.), *Proceedings of the annual research conference of the AIA/ACSA Council on Architectural Research.* Washington, DC: AIA/ACSA Council on architectural Research.

Hassan, I. (1987). *The post-modern turn.* Columbus: Ohio University Press.

Herdeg, K. (1983). *The decorated diagram.* Cambridge, MA: MIT Press.

Hersey, G. (1988). *The lost meaning of classical architecture.* Cambridge, MA: MIT Press.

Hillier, B., & Hanson, J. (1984). *The social logic of space.* New York: Cambridge University Press.

Honour, H. (1968). *Neo-classicism.* New York: Penguin.

Horn, M. (1988, July 18). A new twist on architecture. *U.S. News & World Report,* pp. 40–42.

Ittelson, W. H. (1989). Notes on theory in environment and behavior research. In E. H. Zube & G. T. Moore (Eds.), *Advances in environment, behavior, and design* (Vol. 2, pp. 71–83). New York: Plenum.

Jencks, C. (1977). *The language of post-modern architecture.* New York: Rizzoli.

Jencks, C. (1986). *What is post-modernism?* London: Academy Editions.

Kantrowitz, M., Farbstein, J., & Schermer, B. (1986). The image of post office buildings. In

J. Wineman, R. Barnes, & C. Zimring (Eds.), *The cost of not knowing*. Washington, DC: Environmental Design Research Association.

Kaufmann, E. (1955). *Architecture in the age of reason*. New York: Dover.

Kostof, S. (1985). *A history of architecture: settings and rituals*. New York: Oxford University Press.

Krampen, M. (1979). *Meaning in the urban environment*. London: Pion Limited.

Krier, R. (1979). *Urban space*. New York: Rizzoli.

Krier, R. (1983). *Elements of architecture*. New York: Rizzoli.

Lang, J. (1987). *Crating architectural theory*. New York: Van Nostrand Reinhold.

Laugier, M. (1977). *An essay on architecture*. Los Angeles: Hernessey & Ingalls.

Lawrence, R. (1986). L'espace domestique et la régulation de la vie quotidienne. *Recherches Sociologiques, 17* (1), 147–169.

Lawrence, R. (1989). Structuralist theories in environment–behavior research. In E. H. Zube & G. T. Moore (Eds.), *Advances in environment, behavior, and design* (Vol. 2, pp. 37–70). New York: Plenum.

Le Corbusier (1960). *Towards a new architecture*. New York: Praeger.

Lee, L. -S. (1982). The image of city hall. In P. Bart, A. Chen, & G. Francescato (Eds.), *Knowledge for design* (pp. 310–316). Washington, DC: Environmental Design Research Association.

Lesnikowski, W. (1987). On the changing nature of theories in architecture. *Inland Architect, 31* (5), 28–39.

Macrae-Gibson, G. (1985). *The secret life of buildings: An american mythology for modern architecture*. Cambridge, MA: MIT Press.

Moneo, R. (1978). On typology. *Oppositions, 13*, 23–45.

Moore, C. (1979). *The place of houses*. New York: Holt, Rinehart & Winston.

Moore, G. T. (1988). *Toward a conceptualization of environment-behavior and design theories of the middle range*. Paper presented at the biennial conference of the International Association for the Study of People and their Physical Surroundings, Delft, the Netherlands.

Murray, P. (1978). *Renaissance architecture*. New York: Rizzoli.

Nasar, J. (1983). Adult viewers' preferences in residential scenes. *Environment and Behavior, 15*, 589–614.

Nasar, J. L., & Kang, J. (1989). Symbolic meanings of building style in small suburban offices. In G. Hardy, R. Moore & H. Sanoff (Eds.), *Changing paradigms* (pp. 165–172). Washington, DC: Environmental Design Research Association.

Newman, O. (1980). Whose failure is modern architecture? In B. Mikellides (Ed.), *Architecture for people* (pp. 45–58). New York: Holt, Rinehart & Winston.

Norberg-Shultz, C. (1971). *Baroque architecture*. New York: Abrams.

Norberg-Shultz, C. (1980a). *Genius loci: Towards a phenomenology of architecture*. New York: Rizzoli.

Norberg-Shultz, C. (1980b). *Late baroque and rococo architecture*. New York: Rizzoli.

Norberg-Shultz, C. (1980c). *Meaning in Western architecture*. New York: Rizzoli.

Perez-Gomez, A. (1983). *Architecture and the crisis of modern science*. Cambridge, MA: MIT Press.

Perianez, M. (1984). L'image de l'architecture. *Recherche & Architecture*, No. 59, 11–41.

Popper, K. (1965). *Conjectures and refutations: The growth of scientific knowledge*. New York: Harper & Row.

Prak, N. (1977). *The visual perception of the built environment*. Delft: Delft University Press.

Purcell, A. T. (1984a). The aesthetic experience and mundane reality. In W. Crozier & A. Chapman (Eds.), *Cognitive processes in the perception of art*. Amsterdam: North Holland.

Purcell, A. T. (1984b). The organization of the experience of the built environment. *Environment and Planning B, 11,* 173–192.

Purcell, A. T. (1984c). Multivariate models and the attributes of the experience of the built environment. *Environment and Planning B, 11,* 193–212.

Quatremere de Quincy, A. (1832). *Dictionnaire historique d'architecture.* Paris: Librairie d'Adrien Le Clere.

Rapoport, A., & Kantor, R. E. (1967). Complexity and ambiquity in environmental design. *Journal of the American Institute of Planners, 33,* 210–222.

Relph, E. (1976). *Place and placelessness.* London: Pion.

Robinson, J. (1986). Architecture as cultural artifact: Conception, perception, (deception?). In J. W. Carswell & D. G. Saile (Eds.), *Purposes in built form and culture research: Proceedings of the 1986 International Conference on Built Form & Culture Research* (pp. 98–102). Lawrence: University of Kansas.

Robinson, J. (in press). Premises, premises: Architecture as cultural medium. *Midgaard, 1,* 2.

Robinson, J. (1988). Institution and home: Linking physical characteristics to perceived qualities of housing. In H. van Hoogdalem, N. Prak, T. J. M. van der Voordt, & HB. R. van Wegen (Eds.), *Looking back to the future* (pp. 431–440). Delft, the Netherlands: Delft University Press.

Rochberg-Halton, E. (1986). *Meaning and modernity: Social theory in the pragmatic attitude.* Chicago: University of Chicago Press.

Rossi, A. (1982). *Architecture of the city.* Cambridge, MA: MIT Press.

Rowe, C., & Koetter, F. (1978). *Collage city.* Cambridge, MA: MIT Press.

Sadalla, E. K., & Oxley, D. (1984). The perception of room size. *Environment and Behavior, 16* (3), 394–405.

Schermer, B. (1987). User involvement in aesthetic design decisions. In J. Harvey & D. Henning (Eds.), *Public environments* (pp. 103–108). Washington, DC: Environmental Design Research Association.

Seamon, D. (1982). The phenomenological contribution to environmental psychology. *Journal of Environmental Psychology, 2,* 119–140.

Senkevitch, A. (1983). Aspects of spatial form and perceptual psychology in the doctrine of the rationalist movement in soviet architecture in the 1920s. *Via, 6,* 78–115.

Sime, J. (1986). Creating places or designing spaces? *Journal of Environmental Psychology, 6,* 49–63.

Sixsmith, J. (1983). Comment on "The phenomenological contribution to environmental psychology," D. Seamon. *Journal of Environmental Psychology, 3,* 109–111.

Sixsmith, J. (1986). The meaning of home: An exploratory study of environmental experience. *Journal of Environmental Psychology, 6,* 281–298.

Stokols, D. (Ed.). (1977). *Perspectives on environment and behavior: Theory, research, and applications.* New York: Plenum.

Tuan, Y.-F. (1977). *Space and place: The perspective of experience.* Minneapolis: University of Minnesota Press.

Tzonis, A., & Lefaivre, L. (1986). *Classical architecture: The poetics of order.* Cambridge, MA: MIT Press.

Venturi, R. (1966). *Complexity and contradiction in architecture.* New York: Museum of Modern Art.

Verderber, S., & Moore, G. T. (1977). Building imagery: A comparative study of environmental cognition. *Man–Environment Systems, 7,* 332–341.

Vernez-Moudon, A. (1986). *Built for change: Neighborhood architecture in San Francisco.* Cambridge, MA: MIT Press.

Vernez-Moudon, A. (1987). The research component of typomorphological studies. In R. Shibley, W. Mitchell, W. Pulgram, & R. McCommons (Eds.), *Proceedings of the annual research conference of the AIA/ACSA Council on Architectural Research.* Washington, DC: AIA/ACSA Council on Architectural Research.

Vidler, A. (1977). The idea of type: The transformation of the academic ideal, 1975–1830. *Oppositions 8,* Spring, 95–113.

Vidler, A. (1987). *The writings on the wall.* Princeton, NJ: Princeton University Press.

Wohlwill, J. (1976). Environmental aesthetics: The environment as a source of affect. In I. Altman, & J. F. Wohlwill (Eds.), *Human behavior and environment: Advances in theory and research* (Vol. 1, pp. 37–86). New York: Plenum.

Young, D. (1979). *The interpretation of form: Meanings and ambiguities in contemporary architecture.* Unpublished M.Sc. dissertation, University of Surrey, Guildford, England.

Zevi, B. (1978). *The modern language of architecture.* Seattle: University of Washington Press.

2

Design Theory from an Environment and Behavior Perspective

JON LANG

There has long been an interchange of ideas between the design fields and the sciences. This is generally recognized, but much confusion still exists about the interrelationship between the two. The confusion is unnecessary. In this chapter, there are two interwoven purposes to the review of design theory from an environment–behavior perspective. The first is to clarify the relationship between the sciences and the design fields. The second is to consider the contribution to the design fields that the behavioral sciences have made and can make. A number of issues have to be considered in making this clarification.

Designers and environment–behavior researchers alike need to better understand the design fields. The first issue of concern is the ambiguity of the role of the designer: Is the designer an artist or an environmental designer? The second has to do with the dual consideration of the design fields as disciplines and as professional fields. The third has to do with the nature of the theoretical base for design—what it is, what it can be, and what it should be. Addressing and clarifying the nature of these issues will provide an understanding of the design fields that is imperative for people in the behavioral sciences if they are to make an

Jon Lang • School of Architecture, University of New South Wales, Kensington, N.S.W., Australia.

effective contribution to design and for designers if they are to make effective the potential contribution of the behavioral sciences to their own work.

A look at the history of the modern movements in architecture and urban design and their descendents provides a framework for discussing the shifts in areas of concern and thus the self-image of the designer and the relevancy of the behavioral sciences to design theory. The behavioral sciences have, indeed, for better or worse, made major contributions to design thought. They continue to do so and will do so in the future. The degree to which they do depends on both design ideology and the clarity of the understanding that behavioral scientists have of the design fields, what designers need to know, and how designers use information/knowledge. This chapter thus concludes with a normative statement of the potential contribution of environment–behavior research to the design fields before asking the question: What are the implications for future research and applications?

While this chapter draws on empirical studies, it is not based on empirical research. It is based largely on my own experiences as an educator and as a professional working at the intersect of the design fields and environment–behavior research. I was socialized into the culture of architectural design, so this chapter is really a look at design theory from an environment–behavior perspective by somebody from the design fields.

UNDERSTANDING THE DESIGN FIELDS: THREE CLARIFICATIONS

There is a perception among many current environment–behavior researchers that the design fields, because of their recalcitrance, are unwilling to use the knowledge generated by research. There is a perception by designers that much of current research is irrelevant because it does not address the problems that designers confront in the performance of their professional tasks. The research does not deal with the design fields as professions. Yet, very often, environment–behavior researchers tell designers that the design problems have been incorrectly defined. Many designers, on the other hand, believe that they know better, based on their experiences and common sense. While there is some truth to both positions, many of the reasons for past, and certainly future, opportunity costs associated with building, urban, and landscape designs occur because of incorrect predictions made before they are built of how the designs will work. These predictions are based on the designer's image of people and their natures. The images, in turn, are based on the designer's own experiences and prejudices. The mutual misunderstanding of the concerns of designers and those of behavioral scientists in dealing with design issues unnecessarily maintains this situation.

The Design Fields as Art and as Environmental Design

Architects, urban designers, landscape architects, and city and regional planners have divergent images of themselves and their central concerns. Design professionals have long been rent by two opposing self-images—that of themselves as artists and that of themselves as environmental designers. This is particularly true of architecture at present, but this has not always been so. Architects tend to think of buildings as objects and are thus concerned with object perception rather than environmental perception (Gibson, 1966; Ittelson, 1973). They are concerned with buildings as art rather than environments.

An object can be regarded as a work of art if it communicates a message from one person or group to another person or group. Some objects are purposefully designed to serve this purpose; others acquire this role over time. In the latter case the work may not have been perceived as a work of art originally but acquires this meaning. In this sense, works of art are artificial displays. The display may be a formal one, dealing with pattern/geometry *per se*, or a symbolic one, dealing with associational meanings.

Those architects who think of themselves as artists consider their work to be works of art—a message from a designer to whom it may concern, usually other architects and architectural critics. Those who consider themselves as environmental designers are much more concerned with providing a service to people by improving the quality of the overall human habitat. Few architects would place themselves at the extreme ends of an artist–environmental designer scale, but these are two contrasting self-images, with the former being the one promoted by schools of architecture and the press.

The problem is that few architects or schools of architectural education explicitly recognize the tension between these two self-images. This recognition is essential when considering, from an environment–behavior perspective, theories in the fields of architecture, landscape architecture, and urban design. City and regional planning as a field has clearly come down on the side of environmental design—particularly the nonphysical aspects of the environment (i.e., its social and economic structure). Landscape architecture and urban design recognize more clearly than architecture the duality of the artist–environmental designer image but have had difficulty in coming to grips with it. Architecture has shifted back and forth. It currently sees itself as a fine art concerned with the self expression through built form of the architect's image. While the contribution of environment-behavior researchers to the work of designers as artists—object makers—is important, it is to their work as environmental designers that it is crucial.

THE DESIGN FIELDS AS PROFESSIONS AND AS DISCIPLINES

There is another discrimination which those of us primarily concerned with the education of design professionals need to make more explicit in looking at our own fields. This is the distinction between our fields as professions and our fields as disciplines (Anderson, 1988). For instance, architecture is not only art and/or environmental design but also business. Although the discipline and profession intersect, there is much in the professional activity of designing buildings that falls outside the scope of the discipline of architecture and vice versa. For example, while telling us about architecture as a discipline, much architectural history is not directly and immediately applicable to designing today. Similarly, there is much about the production of buildings that falls within the scope of other disciplines—business management, for instance. This is also true of the contribution of environment–behavior studies. They can make a contribution to both, but much of the contribution is to the discipline of architecture, rather than to the profession *per se.*

THE MEANING OF THEORY

The design fields have a different concept of theory from that of the sciences. It is important for both designers and scientists to recognize this. This is a third area in which clarification is required if designers are to improve their own disciplines and professional actions and, simultaneously, for environment–behavior researchers to recognize their present and potential contributions to the design fields. There have been a number of attempts to achieve this clarification (e.g., Lang, 1980; Moore, 1988).

There are major similarities between designing and doing research. Buildings and theories are both designed. They are based on knowledge, experiences, and intuitions. In this sense buildings are hypotheses—they are predictions about how a set of needs can best be met through changing the physical structure, or layout, of the world. Scientists test their theories by testing hypotheses derived from those theories. Designers and scientists differ in the type of theory on which their hypotheses are based, in the nature of the testing procedures they use, and in the role of theory in their work.

In designing buildings, architects are not directly concerned about generalizing from the results, although they may do so based on casual observations of those results (i.e., their testing procedures are weak). They are concerned with the design and construction of a product—an artifact. One of the basic rules of scientific research is to formulate hypotheses so that generalizations—theories—can be constructed from the results. Another difference between the two is purpose.

The goal of science is to develop theories—descriptions and explanations of phenomena and processes of the world. Architects deal with the future. They rely on their understanding of phenomena to predict the outcomes of their work on whatever aspect of it—art or environmental design—with which they are concerned. The goal of architects is to build well. This has to be done in a changing world. Architectural theories tell architects what others have done and what they should do. They are prescriptive. Often the external validity of these theories is unknown, for they are largely untested statements of faith. The outcomes of their use are seldom rigorously studied. The goal of science is to build good theories—good explanations of phenomena. Designers use theories to guide action—their theories are largely descriptive and prescriptive.

Architectural treatises have long contained statements about the characteristics of good buildings and good environments and the design principles that are required to meet those ends. From Vitruvius (ca. 10 B.C., reprinted 1960), writing two thousand years ago, to Alberti (1755/1955), to the architects of the modern movement (Conrad, 1970), to the postmodernists and deconstructivists of today, architects have been concerned with the outcomes of their work—particularly as perceived by themselves, other architects, and critics. (True, some architects profess not to care about what others think of their work or how well it functions. They serve some higher ideal—art.)

Architectural theory is generally thought to consist of the recording of the changes in appearances of buildings and their spatial organization over time—a description of products. The label "history and theory" is often applied to the discussion. It is seen within an art history framework, not an environmental design one. Explanations are given largely in terms of the creative act. A much more accurate title would be "history *of* theory"—a history of prescriptions and their ideological bases. Should one really call this theory? Yes, *normative* theory, because it deals with the standards, rules, or perceptions of rectitude of a particular group of architects. From very early in architectural history, we have guidelines for design. Writing to Caesar (Augustus, it is generally assumed), Vitruvius wrote: "I have drawn up definite rules for you . . . for in the following books I have disclosed all the principles of the art" (Vitruvius, ca. 10 B.C., reprinted 1960, p. 4).

Drawing on Vitruvius, the writings of Alberti (1404–1472), Serlio (1475–1554), and Palladio (1508–1580) are in a similar vein. They are concerned with the number, arrangement, and measure of the elements of architecture drawing from what they perceived had stood the test of time. They were concerned with the appearance of the geometry of buildings—with establishing canons of beauty—and with a salubrious environment for life.

Designers do not always do what they profess. There are a number of possible reasons for this. The context may not allow them to do so, the conditions may be restrictive, or their clients may be politically powerful (see also Lang, 1988b). Thus one has to distinguish between what designers say and what they do, between professed and practiced normative positions.

Normative theories are always based on some intellectual construct of how the world works and attitudes toward what it is. The constructs, or models, are assertions about reality—what things are and why. Borrowing from fifth-grade English grammar classes and the work of others, I have called these constructs "positive theory" (Lang, 1987). Architecture, too, traditionally has had a positive theoretical base. It has been largely intuitively derived and has been poorly articulated. Nevertheless, it has worked well as the basis for normative theories by accurately predicting outcomes of designs in societies with a shared value system, where change occurred slowly. It still works in such societies. In complex, changing societies, relying on intuitively built positive theoretical bases has often led to very misleading predictions about what design outcomes will be—especially in terms other than the architect's own.

Architectural theories, both positive and normative, have been influenced by contemporaneous philosophical ideas. They have always been influenced by concepts of the human being and of nature—concepts of environment and behavior. This has been especially true when a strong environmental design attitude has persisted, but also when the design fields have focused on themselves as art.

Architectural theory has largely focused on products—built form. Until recently there has been very little explicit concern with the design process. It has been *substantive* in nature, not *procedural*. The process was regarded as something carried out intuitively and magically, given a program/brief by the client. Historically there has been very little concern for the nature of the program and the architect's responsibilities in its development. The statements on the nature of the program which have been expressed in the architectural literature from the beginning of the nineteenth century, clearly show that the program was something generated by the client. Programs were based on the client's experience, however limited it might be. The program was largely considered in terms of rooms required and, perhaps, the connections between them. The architect's task was to give form to these. Aesthetic objectives were not part of the program. They were, and still are, usually regarded as the prerogative of the architect. One of the major contributions of the behavioral sciences has been the introduction of a greater conceptual clarity in the programming process. This has been done through the use of models of behavior–environment interaction and methods/procedures for obtaining design information and evaluating designs.

One of the major developments in the discipline of architecture during the past three decades has been the development of *procedural theory*—design methodology. Procedural theory also has positive and normative components. The goal of positive theory is to describe and explain the overall decision-making process, how its phases are carried out, and why they are carried out in this way. It is also to understand the implications of doing design in one way rather than another. It is not a well-developed science, but it can be. Normative procedural theories consist of ideological statements of how the process should be carried out. Implicit, if not explicit, in the writings of architects from Alberti (1755/1955) to Le Corbusier (Eslami, 1985) are statements on design methods. However, the systematic consideration of the design process is still missing from architectural educational curricula.

Theory for the design fields can thus be said to consist of four overlapping bodies of knowledge: *positive* and *normative* statements on *substantive theory* that deal with the nature of the built world and *positive* and *normative* statements on *procedural theory*—design processes. This can be displayed in a two-dimensional matrix (see Figure 1). Environment and behavior studies are concerned with developing positive substantive theory—descriptions and explanations of the phenomena with which architects deal and also methods of, at least, programming and evaluating in carrying out the design process. Certainly there has been a proliferation of books on this subject (e.g., Palmer, 1981; Peña, 1977; Preiser 1978; see also Sanoff, 1989).

Much remains to be done in building such a theoretical basis for the design disciplines. A look at the design fields shows that most of what is subsumed under the title "theory" is normative rather than positive, and substantive rather than procedural. The architectural movements of this century can be seen as exemplars of this.

Subject Matter of Theory	Orientation of Theory	
	Positive	Normative
Procedural		Professed
		Practiced
Substantive		Professed
		Practiced

FIGURE 1. A quadrapartite model of design theory.

THE MODERN MOVEMENT AS ART AND AS ENVIRONMENTAL DESIGN

The term "modern" has been used to describe particular architectural movements and styles since the beginning of the Renaissance. Today we think of modern architecture and urban design as that which flourished during the first six decades of this century, but its roots can clearly be traced back to the middle of the eighteenth century (Collins, 1965). The more narrow view will be taken here, because it is during the last hundred years that the systematic, although sporadic, studies on environment and behavior have been carried out. The concern here is with the period in architectural history beginning with the last decade of the nineteenth century and lasting until the 1960s—a period of major technological and social changes in the world. It ends with the recognition of the limitations of modern design theory and particularly with its positive theoretical base.

There are a number of ways of identifying and categorizing the various streams of thought that come under the rubric of the modern movement. The categorization used here is by no means universally accepted. However, it shows well the shifts in consideration of architecture between environmental design and art.

THE FIRST GENERATION: THE ANGLO-AMERICANS AND THE CONTINENTALS

The first generation of the modern movement consisted of two contemporaneous groups of architects. One group has been called the "Anglo-Americans," the "decentralists," or the "regressive utopians" because they looked to the past for their models of good societies and toward rural imagery for their aesthetic. The other group has been called variously the "Continentals," the "centralists," or the "progressive utopians." They looked forward to the harnessing of technology to raise the standard of living of all people. They also formed a new aesthetic philosophy without explicit historical precedents.

The two groups had much in common. They were both very much concerned with environmental design—with the creation of better habitats for all people. They were concerned not only with the whole social and moral organization of society but with its aesthetic expression as well. However, their images of the good life and consequent proposals and designs were very different. They were based on different normative models of human behavior and of the environment. Within each school the members were diverse, but some of the ideas and work exemplifies the thinking.

Exemplifying the Anglo-American group are the works of Ebenezer Howard, Clarence Stein and Henry Wright, and Frank Lloyd Wright in terms of urban design and the last mentioned also in terms of architecture (see, for example, Howard, 1902; Stein, 1951; Wright 1958, 1960). Radburn, New Jersey, is a product of this line of thought (see Figure 2). The Continental school consisted of two major branches in terms of its influence on the world: (1) that associated with Le Corbusier and CIAM (Congrès Internationaux d'Architecure Moderne; see, for example, Le Corbusier, 1973, and Figure 3); (2) those associated with the Bauhaus and its antecedents in the rationalist and constructivist schools of the postrevolutionary Soviet Union (Wingler, 1969).

The Anglo-Americans and the Continentals shared many observations about the cities of Europe and North America. They both deplored the living conditions of the poor in the large industrial cities. They were both concerned with the alienation of people from the natural environment, the long daily journeys of many workers, the inefficiency of the circulation framework of cities, the automobile, pollution, the unjust social stratification of people, the lack of light, sunlight, and ventilation in much housing, and, more broadly, the overcrowded conditions of cities. Their normative positions also were similar in many ways. They both advocated efficiency in movement systems and the strong segregation of land uses. A common focus of attention was certainly on environmental design—improving the lives of people. They were also concerned with design as art. While the Continental school had a radical aesthetic philosophy, the Anglo-Americans harkened back to the arts and crafts movement of William Morris.

FIGURE 2. Radburn, New Jersey: (a) Plan. From C. Stein, *Towards new towns for America*, New York: Van Nostrand Reinhold, 1951. Reprinted by permission.

The solutions proposed by the two schools differed because they were based on different normative models of the social organization of society and of aesthetics. They differed radically in attitudes toward property ownership, nature, and the social stratification of society. While both schools had roots in the social and philanthropic movements of the nineteenth century, the political roots of the Anglo-American schools were embedded in concepts of Jeffersonian democracy and the aesthetic philosophies of William Morris. The American component was more particularly concerned than their British counterparts with private property rights. However, both groups shared an advocacy for a greater focus on communally held common open space/parks than the prevalent attitudes of governments of the time. The Continentals, influenced

FIGURE 2 (*cont.*) (b) A cul-de-sac today.

by socialist philosophies and the Communist ideals of Marx and Engels, promoted an egalitarian view of society and of communal property ownership.

In their urban design proposals, the Anglo-Americans tended to strive for a revival of the small town of small-scale buildings, residential areas of single family homes with private gardens, and easy access to shops, schools, and parks (Stein, 1951). Their concept of the good life and a good environment remains an important one in architectural philosophy (see Alexander, Ishikawa, & Silverstein, 1977). The aesthetic philosophies of individual architects varied considerably. Many of the Anglo-American architects looked back to rural architectural models but there are major differences between the architecture of Frank Lloyd Wright and Henry Wright. While both were very much influenced by the rural environment, Frank Lloyd Wright's own aesthetic philosophy, and that of his clients, was more progressive (see Eaton, 1969).

The Continentals prescribed tall buildings, homogeneous in character, in parklike settings. Their emphasis was on the healthful environment full of open space, active recreational opportunities, and public

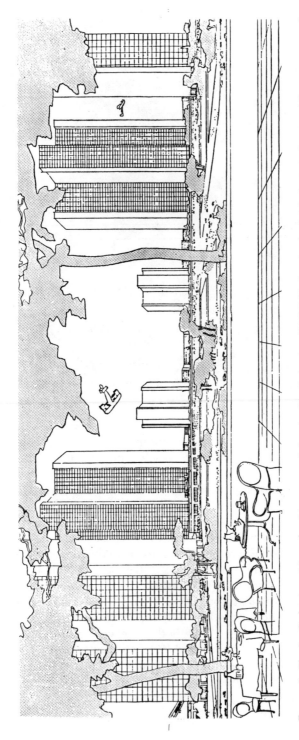

FIGURE 3. The Le Corbusian vision of architecture and the city (Le Corbusier, 1934/1967; drawing by Le Corbusier). Reprinted by permission of Fondation Le Corbusier and VAGA, New York.

ownership of real estate. Their aesthetic philosophies, in strong contrast to the Anglo-Americans, owed much to the ferment in the art world and particularly to the emergence of cubism. The clients of the single family homes that they designed tended to be members of the avant-garde in art (see Eaton, 1969).

On the environmental design side, many of the ideas of the first generation that had developed during the first two or three decades of the century had a major impact during the 1950s and 1960s. The Continental school saw its model of the environment applied to urban renewal and public housing schemes across the world in countries as diverse as the Soviet Union and the United States, Scotland and Venezuela, Japan and India. Brasilia is the movement's crowning glory, with Chandigarh not far behind. The Anglo-American philosophy first saw implementation in the garden cities of Letchworth and Welwyn in England in the 1920s and 1930s. It was, however, used much more extensively in a host of post-World War II new towns and suburban developments in the British Isles, and across the world from Scandinavia to the tropics. Columbia, Maryland, and Jonathan, Minnesota, are examples of this line of thinking in the USA.

THE SECOND GENERATION: THE REDISCOVERY OF ARCHITECTURAL SYMBOLISM

Implicit in the thinking of the first-generation architects of the modern movement was a concern with the symbolism of architecture. It tended to be a by-product of other concerns, so their architecture was a symbol of progress: the new vision of a new social order. The architects of the second generation came down strongly on the art side of the art–environmental design continuum. Their architecture is often regarded as a baroque revival in which a sense of order and hierarchy was sought. A concern with symbolic aesthetics as a goal rather than as a by-product of other design decisions came to the forefront. Their striving for monumentality called into question the international style of the Bauhaus, in particular. The lyricism of their designs sought to create emotional space, with architecture being used as a medium of expression.

In many ways this line of thinking was related to the spiritualism of Rudolf Steiner, the founder of anthroposophy, and the theosophy of Madam Blavatsky and Annie Besant. They believed that the method for generalizing knowledge is intuitive thinking—science only gives abstract knowledge of some uniform relations in nature. It is exemplified by such work as Hans Sharoun's Philharmonie in Berlin, Jorn Utzon's

Sydney Opera House, Saarinen's Dulles Airport building in suburban Washington, and Ieoh Ming Pei's Des Moines Art Center (Knevitt, 1985). The arts center might (with its references to the past) be regarded as a prelude to postmodernism. The second generation also includes the work of a number of architects whose early work was part of the first generation. For instance, Le Corbusier's chapel at Notre Dame du Haut in Ronchamp (see Figure 4) shows a major departure from his early work, particularly in its plasticity of form. In form it is more closely related to his abstract paintings than to his early architectural work.

If one considers the focus of the second generation's work in terms of Maslow's model of human needs, the primary concern is with biological needs in terms of the users of buildings and symbolic aesthetics in the architect's terms. One sees little concern for environmental design issues—with the overall human habitat—in the work of architects of this generation. The result was the emergence of urban design as a fledgling field and a shift in the focus of landscape architecture from gardens to broader environmental concerns. The latter had earlier been the purview of architecture.

FIGURE 4. Notre Dame du Haut, Ronchamp, Le Corbusier, architect.

THE THIRD GENERATION: WORKING A FUNCTION INTO A FORM

A third generation of architects began to think of form independently of the traditional, limited concepts of function that were inherited from the Continental branch of the first-generation architects. The process was to take a form and to work a function into it. In this sense the work of these architects harkens back to formal models of eighteenth-century French neoclassical architecture. Louis Kahn is the architect most closely associated with this line of architectural thought in his work for the Salk Institute and the Richards Memorial Laboratories (see Figure 5). Kahn noted that a building can only tell a story about itself when it is not longer in use because then there is no preoccupation with its function (Wines, 1985)! In a somewhat contradictory line of thinking, the concern was also with the reestablishment of a regional architecture—a local identity.

In urban design, the third generation also broke away from CIAM and its manifesto, the Athens Charter. They broadened their concern with the nature of people and cities. A diverse group of architects were identified with this move. Clustered under the label "Team 10" they sought a richer environment than that generated under the principles of the Athens Charter (Smithson, 1968). Yet the environments they designed were just as sterile. They did turn to the behavioral sciences for an understanding of human behavior patterns, but there was little understanding of how the built environment afforded such patterns—little understanding of the environment–behavior interface. Thus, while the Smithsons (1967) were very much aware of the importance of the street in the lives of low-income people in London as portrayed by Michael Young and Peter Wilmott (1957), they were unable to turn that portrayal into a normative design statement in which the essence of the role of the street was retained (see Figure 6).

THE POSITIVE BASIS OF NORMATIVE
MODERN DESIGN THEORY

When one looks at the theories of the modern movement using the quadrapartite model of theory shown in Figure 1 (i.e., positive–normative, substantive–procedural), it is clear that the concern is with producing design principles for physical form (i.e., with normative, substantive theory). Yet all these ideas had a positive basis.

The positive basis is drawn from the personal experiences of archi-

FIGURE 5. Richards Memorial Laboratories, Philadelphia, Louis Kahn, architect.

tects, their attitudes toward the behavior of other people, and prevailing philosophies of the good life in a highly class-oriented and culture-bound manner. Like Le Corbusier, modern architects designed very much for people like themselves. A workaholic, Le Corbusier politically flirted with Communism. He was very much concerned with exercising

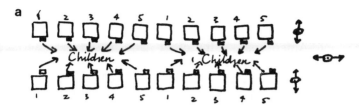

'Cars, telephones,—represent not so much a newer and higher standard of life as a means of clinging to something of the old. Where you could walk to your enjoyment, you did not need a car.' Family and Kinship in East London, Young and Wilmot.

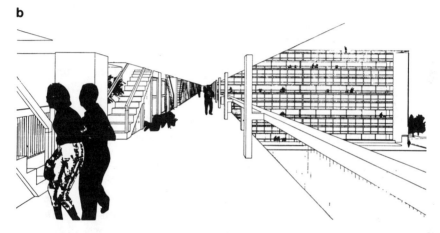

FIGURE 6. The street: (a) The Smithsons' observations of street life. (b) The street recreated. From A. Smithson & P. Smithson, *Urban Structuring*, New York: Van Nostrand Reinhold/Studio Vista. Reprinted by permission.

the body as well as the mind, and his urban schemes reflected this view of life. In looking at the appalling conditions of the industrial city, he could only see the negative and the inefficiencies. He admired the efficiency of the products of engineers and sought the same efficiency in the design of the built environment. Thus, he was concerned with movement channels and the interrelationships of land uses and, thus, with shortening circulation distances.

In the development of his aesthetic philosophy, in his view of architectural form as art, Le Corbusier was very much influenced by both cubism and the aesthetics of machinery. In this he was unlike the architects of the Bauhaus, who favored simplicity, an aesthetic goal derived from, or justified by, the Bauhaus architects' own interpretations of Gestalt theories of perception (see Senkevitch, 1974; Tauber, 1973). This link is particularly evident in the art theories of Kandinsky (see Overy, 1969;

Tauber, 1973). Le Corbusier drew different justifications from it (Gray, 1953).

The Anglo-Americans, on the other hand, based their normative theories initially on beliefs about small-town life (e.g., Howard, 1902) and the sociology of the Chicago School (e.g., Cooley, 1909; see also Michelson, 1976). They also relied on their own experiences and the broader myths of the culture of which they were a part and its antiurban biases (White & White, 1964).

By the early 1960s, the limitations of the modernist model of human beings and claims about human behavior, explicit or implicit, in the work of architects became clear. Buildings, neighborhoods, and city centers simply did not work either instrumentally or aesthetically as intended and predicted. The ambition to create a new morally correct world was challenged by critics such as Jane Jacobs (1961). She argued for greater diversity and richness in design, to reflect the richness and diversity of people. The modernists, in response, argued for their work as a contribution to art, not directly to human life but rather to human life through art. The question remains "Who understands it?"

THE DESIGN FIELDS AS ART

When designing individual buildings—objects in space—and the handling of space within buildings, most of the architects of the modern movement wanted to align themselves with artists. This was also true of the clients who hired them (Eaton, 1969). Thus, Le Corbusier's early clients were clearly supporters of the modern art movements of the first three decades of the century, while Frank Lloyd Wright's had a more conservative view of art (Eaton, 1969).

The goal of design was to communicate meaning through line and form. Artists and critics were well aware of the contemporary writings on the psychology of perception. One of the first positive theories of aesthetics drawing on contemporary psychological theories was put forth in *The Sense of Beauty* by George Santayana (1896/1955). This book had both a direct and indirect effect on design theory. Santayana's concept of beauty as the objectification of pleasure served as the theoretical base for books on design (e.g., Hubbard & Kimball, 1917), but it also inspired research on psychology that later affected design theory. Santayana attempted to describe and explain the nature of the aesthetic experience rather than the nature of the aesthetic principles used by artists. He relied very heavily on the psychological theories of William James (1890). Another of his contemporaries at Harvard was Hugo

Munsterberg, brought from Germany by James. Munsterberg was very much concerned with establishing a positive theory of aesthetics. His theories were an interim step between empiricist and Gestalt theories of perception. They were used extensively by the Constructivist and Rationalist schools of Soviet architecture (Khan-Mohomedov, 1971; Senkevitch, 1974). The reliance of the Bauhaus masters on Gestalt theories of perception developed by Koffka, Kohler, and Wertheimer has been well documented (e.g., Overy, 1966; Tauber, 1973). Cubism received its positive justification in the writings of Charles Henry, whose theory was thus presented in the art journal *L'Esprit Nouveau* (Gray, 1953). Henry perceived a possible correlation between the line and form *per se* of drawings and paintings, and people's emotional responses to them.

These positive theories of perception and aesthetics were used in two ways by architects in presenting their normative theories. Some architects, such as Nikolai Ladovsky at the VKUTEMAS (State Higher Art and Technical Studios), were concerned with building an ideology based on the nature of perception. Others (e.g., the cubists) were concerned with justifying their own intuitively derived positions in terms of the positive theories of the time. They did not recognize the tentativeness with which these theories were presented. This seems to have been the way the cubists approached their theory-building efforts. Henry's establishment of a possible linkage between the line and color of drawings and the emotional responses of the viewer, even though hesitatingly presented, justified the aspirations of abstract artists and much of the thinking of the second-generation architects of the modern movement.

The flow of theories, both positive and normative, across the Atlantic has only been partially documented. A tentative effort to show the exchange is shown in Figure 7. This diagram clearly shows that many architects involved in the development of normative aesthetic theories were concerned with contemporary psychological theory.

THE DESIGN FIELDS AS ENVIRONMENTAL DESIGN

The social and philanthropic movements of the nineteenth century were very much concerned with the quality of life of people, particularly the working classes—groups who were not the patrons of architecture as art. The movements were strong advocates for better working conditions, for the reduction of child labor, for a more salubrious environment, for access to open air, and to the betterment of the mind. Implicit in their work was some concept of a good social organization of society

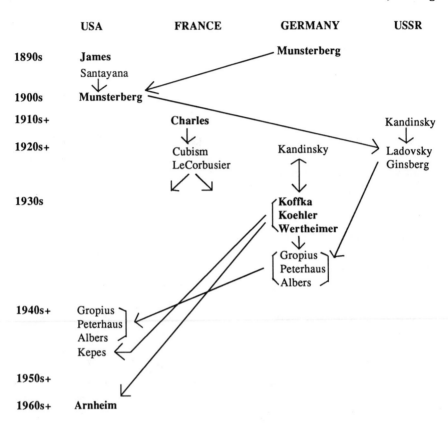

FIGURE 7. The flow of perceptual concepts between psychologists and artists. Psychologists' names are in bold print.

and a good environment in which this should take place. New towns (usually company towns), housing developments, and social facilities resulted from the work of people such as Robert Owen, Octavia Hill, Titus Salt, Jean Baptiste Godin, and Jane Adams and those associated with Hull House and the Settlement House movement. These were considerably better places in which to work and live than the existing industrial towns (Benevolo, 1971). This concern with a good living environment carried through to the first generation of the modern movement (Scott Brown, 1976).

It is particularly clear with the first generation of modernists that a prime consideration in their manifestoes was with the nature of the environment and the nature of human behavior, along with questions of what makes a good city or a good neighborhood—good people and a good life. Their ideal schemes were powerful, if oversimplified, models of human existence and aspirations. They were highly normative in character based on images of how the designers felt people should live. The conceptual diagrams and designs that resulted have left a powerful imprint on our minds: the Garden City of Ebenezer Howard (see Figure 8), the Industrial City of Tony Garnier, the Radiant City of Le Corbuiser (see Figure 9), the neighborhood unit of Clarence Stein (see Figure 10). They were all built on a concept of the family, its members, their activities, and of a social organization knitting communities together.

In architectural terms, an underlying theme was also a concern for the character of enclosed and open space and the human experience of the sequence of spaces. Indeed, it is the attitude toward spatial organization that distinguishes the modern movement from its predecessors (Gideon, 1941). Yet the space was seen as an object of artistic creation as much as for the housing of an activity. The artistic component—an expression of the architect's image—took precedence over the behavioral component—the nature of the experience of the people who use the space (environment). Nevertheless, implicit in the design was a model of human experience. It was a biological and social model—rich on the biological side and weak on the sociological and psychological.

The second generation, while continuing its concern with the spatial organization of buildings, was less concerned with their social purpose. It was concerned with environmental design on a small scale—in the design of interiors of buildings, but not in the broader human habitat. The third generation only half-heartedly addressed the issues of the living environment. While it is true that Team 10 was concerned with housing and cities and was indeed influenced by the work of sociologists, it made little progress on the ideas of the first generation.

It is clear that Le Corbusier was very concerned with human beings and their lives. Yet he had a limited and standard model of the human being. Although he paid lip service to the diversity of human lifestyles and aesthetic values, he focused on the similarities among people (Le Corbusier, 1923/1970; 1951/1968). The model of the human being used by the Anglo-American stream was a broader one, and the resulting architecture "less bold." Thus, it was, and is, held in lower esteem by the design professions.

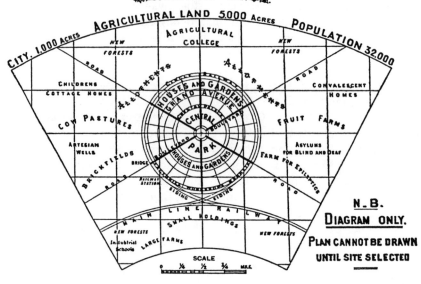

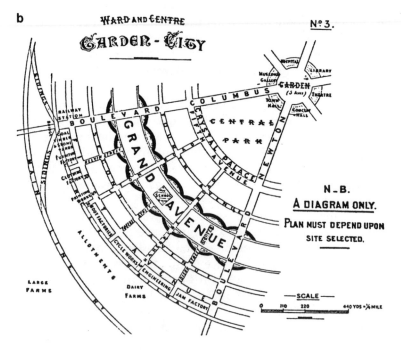

FIGURE 8. Garden City: (a) Garden City and rural belt. (b) Ward and center of Garden City. From E. Howard, 1902.

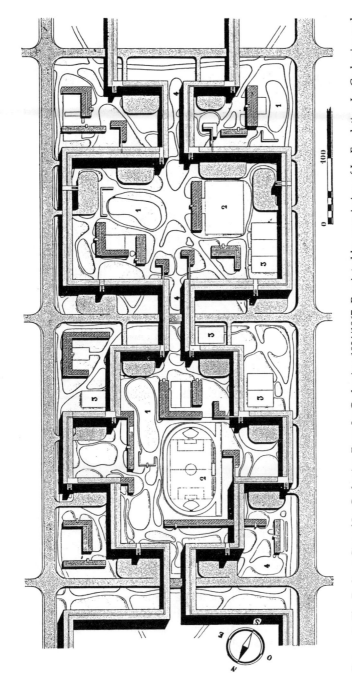

FIGURE 9. The Radiant City of Le Corbusier. From Le Corbusier, 1934/1967. Reprinted by permission of Le Fondation Le Corbusier and VAGA, New York.

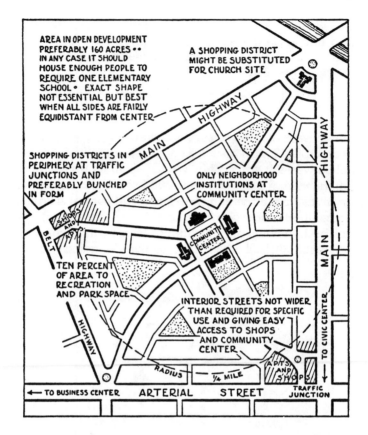

FIGURE 10. The neighborhood unit. From New York Regional Survey, 1927.

ENVIRONMENT–BEHAVIOR STUDIES AND MODERNIST THEORY: A COMMENTARY

If one reviews the manifestoes of the modernists, their observations of how the world works and the nature of the person–environment interface, one is now, with the wisdom of hindsight, struck by the naïveté of their ideas. Environment–behavior studies *per se*, on which design theories and proposals could be built, were unknown. What has been described above are sporadic efforts by architects to inform themselves by looking at contemporary social and behavioral science theories. The knowledge they gleaned never became part of their own theoretical base. A look at their work in terms of the model of theory introduced above illustrates this. This discussion is thus divided into two interrelated parts: substantive and procedural.

SUBSTANTIVE THEORY

In dealing with people, environment, and behavior, it is not surprising that the architects of the first half of this century relied so much on their intuitions and common sense. It is also not surprising that they relied heavily on political agendas in establishing ideas about what is a good environment. They were moral absolutists. They were, however, willing to learn and did learn from what might be regarded as the predecessor of what we call "environment–behavior studies." Yet these studies were in their infancy.

It was only with the blooming of research and theory building since World War II and the obvious dissatisfaction of the inhabitants of many new buildings and urban and city designs that the limitations of the thinking of the modernists became clear. The criticism has been well documented, and only an overview will be given here (see Blake, 1974; Brolin, 1976; Jacobs, 1961; Wolfe, 1981). What is interesting, however, is the response of architectural theorists and the conflicts in opinions about the future direction for the discipline and the profession.

The criticism of the modern movement boils down to a few simple and interrelated statements. A number of the models on which architects were relying as a basis for design were too simple. The model of the environment was too simple, the model of people was too simple, and the model of the interrelationship between person and environment was not only too simple, but also largely erroneous.

The environment was regarded largely as a physical setting for human behavior. The surroundings of people are more complex than that. Our surroundings also consist of a cultural and social dimension. Further, the physical, cultural, and social contexts are inextricably linked in an open, not a closed, system.

People are diverse, too, with a range of human needs well beyond those considered by designers. Much of the modern movement's philosophy was based on a consideration of only two of the basic human needs identified by Abraham Maslow: the physiological and aesthetic. Little concern was paid to needs for affiliation, esteem, or self-actualization (Belgasem, 1987). It is true that CIAM broke up when it tried to come to grips with these more complex needs. Simple diagrams of urban form neither allowed for nor afforded them. Little attention was paid to the behavioral forms (i.e., manifestations) of these needs within different cultural or terrestrial contexts (Brolin, 1976; Rapoport, 1977) or the architectural consequence.

Assumptions about the nature of the fundamental human behavioral processes—human motivations, perception, cognition and affect, and spatial behavior—were limited. Limited attention has been paid to the multi-modality of perception and to the role of movement in veridical perception and thus the experiencing of the world (see Rasmussen, 1959, on the modality of "experiencing architecture"; Thiel, 1961, and Cullen, 1962, on the role of movement; and Gibson, 1950, 1979, for the psychological theory). New developments in dissonance theory (see Festinger, 1957, for its original formulation) and the formation of human attitudes have given us a much greater appreciation of the role of individual experience in establishing environmental preferences. In dealing with spatial behavior, too little attention was paid to the concepts of privacy and community and the mechanisms for achieving them. We now know better (see Altman, 1975; Newman, 1979). It was shallowly assumed that communal organizations could be designed and implemented self-consciously (see Gottschalk, 1975).

In dealing with the human–environment interaction, it was assumed that the physical environment and, in particular, the artificial architectural component of it had a deterministic effect on many aspects of human life: spatial behavior, social organization formation, and aesthetic experience. The environment was seen as the stimulus and behavior the response, without much consideration for intervening variables. This is a naive model of the behavior–environment interface. It is still the predominant one in architectural thought. The role of social and cultural aspects of the environment, including political and economic aspects, were neglected as factors in the design of buildings, landscapes, and urban layouts.

PROCEDURAL THEORY

Designers have some intuitive understanding of the overall design process. All designers follow some normative procedure in their work. Attempts to describe and explain the process in positive theoretical terms have been sporadic. Vitruvius and Alberti described the process as one of selecting parts; Descartes described it as one of decomposing a problem, solving the subproblems, and composing these solutions into an overall answer. The modernists took this no further.

There are few clearly articulated normative models emerging from the modern movement. The models that exist are based on Descartes' concept of design as a "rational" process. Le Corbusier described his own process in this way in *Vers une architecture* (1923/1970). His model can be described in a number of steps: the formulation of the problem in terms of the activities to be housed, the formulation (design) and application of design standards, and the intuitive integration of these into a work of art. The sequence of activities was largely internalized and undifferentiated. This is a positive statement. It was also a normative statement for the designers of the modern movement.

The "failure" of the modern movement has led to an increased inquiry on the nature of the design process, its overall structure and design methods. This effort can be traced to the 1960s, when a number of publications and conferences initiated the exchange of ideas on procedural theory (see Jones & Thornley 1963; Moore 1970). These have paved the way for more comprehensive statements on the process (e.g., Broadbent, 1973; Koberg & Bagnall, 1974; P. Rowe, 1987; Zeisel, 1981). There is still much that we do not understand.

MODERNIST THEORY IN PERSPECTIVE

The theories of the designers whose work falls under the rubric of the modern movement consisted of (1) a program, in abstract ways, of a social and political organization of society, and (2) an aesthetic philosophy. CIAM and the Team 10 sought not only to persuade but also to recruit adherents, demanding a "commitment to architecture." While addressing a wide public, the theory sought to confer some special role of leadership on a set of designers.

In this enterprise, designers sought to find certainty in architectural history and their own experiences. Their approach was more in the field of formal design than on the relationship between economic, social, and aesthetic values. Much of the result is architecture without life—architecture as a *symbol* of architecture. This was not because designers were not well-meaning, but because they were relying on a self-referent body of positive knowledge and images of the good life drawn not from life but from a "religious" ideology. They were still imbued with the *esprit des Beaux Arts*. Based on their own observations and values, they designed from the top—the intelligensia—to the bottom—the vulgar people (see Brolin, 1976). Not much has changed.

OUR CONTEMPORARY DESIGN THEORIES

We live in a time when the dominance of Christianity and so-called Western culture has been challenged. Gender roles and the role of the family in society are changing. The elite-based culture is being challenged (Foster, 1985).

There have been two directions of response to the observations of the limitations of the normative theories of the modern movement. The first has been to shift to new normative positions without shifting considerations of the nature of the positive base for design. The second has been much more concerned with the development of an understanding of the experiential nature of the human–environment interaction—with a normative model of design based on an environment–behavior approach. The first direction is the dominant move in architecture today—a continuation of the philosophy of architect as artist and architecture as an art form. The second, with its concern for a broad environmental design, is in its infancy. The former receives most recognition. Today's architectural heroes are Isozaki, Jahn, Gehry, Graves, Meier, Moore, Stirling, and Venturi—with young architects adding Botta and Rossi to their lists (Dixon, 1988). The latter approach is associated most closely with the work of Christopher Alexander (Alexander *et al.*, 1977). However, even he takes what anthropologists call an *etic* approach (Vernez-Moudon, 1988). While concerned with how people experience the environment it is he, himself, who states the normative position. In contrast, the work of people like Amos Rapoport (1977) is developing the positive side of substantive design theory, but no clearly articulated normative theory has yet emerged.

POSTMODERNISM, DECONSTRUCTIVISM, AND CLASSICISM:
ARCHITECTURE AS ART

Three major normative theories have replaced modernism in recent years: postmodernism (see Figures 11 and 12), deconstructivism (see Figure 13), and classicism (see Figure 14). They stand for architecture as high art. While their promulgators would probably dispute it, these movements represent a major withdrawal from the consideration of architecture as environmental design.

The term "postmodern," while meaningless in standard English, is applied to a number of movements that emerged in the 1970s. The distinguishing goal of postmodernism was/is to create architectural patterns that have dual referents—ambiguity. Postmodernism pays attention to historical memories of building types and styles. The mechanism is to incorporate past building and decorative elements that are understood on one level by other architects and the cognoscenti (i.e., patrons of art) and on another level by the lay public. The goal was to eliminate the "alleged elitism of rational imagery" (Eisenmann, 1982; Klotz, 1989). In its

FIGURE 11. The Guild House, Philadelphia, Robert Venturi, architect: An early example of postmodernism.

purest sense postmodern refers to the use by architects of elements from the history of architecture, either directly or as reinterpreted. More loosely, it refers to any building that differs from the modernist aesthetic. Postmodernism has held the center of attention as the prevailing normative theory amongst the architectural intelligensia from the 1970s until recently giving way to deconstructivism.

FIGURE 12. Recent postmodern architecture—housing, Tegel, West Berlin, Charles Moore, architect.

Postmodernism is derived from modernism and shares much of its philosophical underpinnings. It can be seen as an extension of the second and third generations of modernism, with its focus on symbolic aesthetics and with the role of buildings as communicators of meaning. It is a continuation of the second generation's search for symbols through expression and the third generation's concern with regionalism. What is different is postmodernism's reliance on associational meanings. Robert Venturi's *Complexity and Contradiction in Architecture* (1966) recognized architecture as a means of discourse in which meaning is conveyed by familiar signs and icons.

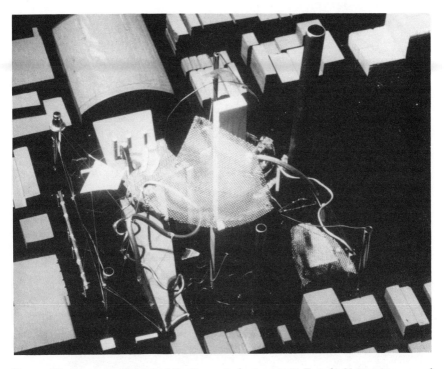

FIGURE 13. Deconstructivist architecture—student project, Temple University, second year—Dilger Studio.

Postmodernism also means a challenge to familiar rules and conventions of architectural design. Architects such as Venturi, Robert Stern, Michael Graves, and Charles Moore believed that the language of modern architecture was too abstract. They introduced populist elements into their own languages, but this was still an etic approach. The decision of what would communicate to the lay population was largely intuitive. The result has been that the predictions or claims of what is communicated are largely erroneous (Groat & Canter, 1979). In this sense, the attitude toward theory and theory building is the same as that of the modernists. The results are visually dramatic (Jencks, 1986). The designs are richer and more embellished than those of the modernists. The view was still, however, with architecture as building in space—in the way architecture is photographed.

Developed from the deconstructionist philosophy of Jacques Derrida, deconstructivism is a current "architecture as art" normative theory that is part of a larger cultural direction of society—punk fashions

FIGURE 14. Contemporary classicism, John Blatteau, architect.

and rap music. It is a way of reading paintings and buildings as text in which conflicting meanings undermine fixed interpretations. It also owes much to the latter phases of the constructivist movement in the Soviet Union. It was first applied to the analysis of the works of artists such as Cy Twombly and Alselm Kiefer. Architects such as Zaha Hadid find their aesthetic inspiration in a sense of disruption, dislocation, and deflection (Jencks, 1988; Cornwell, 1989; Norris & Benjamin, 1989).

It is very much a formalist view of architecture. This view has evolved from postmodernism in taking fragmentary images out of their original context, combining such things as twentieth-century pop themes with fifteenth-century Renaissance or baroque elements. The goal is to undermine traditional expectations of forms; deconstructivism takes delight in reversing the normal grounds for meaning. Charles Jencks (1988) suggests that it should be seen as a "joke made from within the system and told with a wink." This supports Vernez-Moudon's view of our contemporary theories as very much taking an etic rather than an emic approach (Vernez-Moudon, 1988).

Like other broad labels given to architectural design movements, deconstructivism embraces a fairly diverse set of architects and by no means all their work. In Europe we have people like Rem Kolhaas, who seeks an indeterminancy and minimal coherence in design, and groups such as Morphosis. In the United States there is a variety of work by people such as Lars Larup, who sets out to show the arbitrariness of cultural forms; Frank Gehry, who assaults the conventions of Modernism and building type; Daniel Libeskind; the work of SITE, which is set around themes of disaster; and some of the work of Peter Eisenman, which seeks to alienate the world (Cornwell, 1989; Johnson & Wigley, 1988).

Paralleling the development of postmodernism and deconstructivism has been the revival of classical architecture. Pursued by Leon Krier in Continental Europe, Robert Adam and Quilan Terry in the United Kingdom, and Alan Greenberg and John Blatteau in the United States, it argues that the classical vocabulary of architecture is a timeless communicator of meaning. Other architects take a more adaptive use of classical design principles and fall somewhere between postmodernism and pure classicism (e.g., R. A. M. Stern, 1988).

Community Architecture: Architecture as Environmental Design?

Another new development in normative design theory is the community design movement, which focuses on a community's participation in the decisions that affect its life—an emic approach to normative theory. It has been primarily concerned with residential communities, but its logic can be and has been extended to educational and commercial communities. It grew out of the activist spirit of the 1960s, but also from the strong criticism of (1) the attempt to solve urban problems through physical design and (2) the elitist values upon which concepts of good communities were built. This criticism was led by people such as Paul Davidoff and Herbert Gans, both social scientists teaching in a school of design (Scott Brown, 1976). The movement led to the development of the whole field of environment–behavior research.

Often perceived as something which grew and died in the 1960s, the community design movement persists and has been given strong royal support in the United Kingdom by the Prince of Wales (Holden, 1988). As such it persists most strongly in the United Kingdom and Continental Europe. English architects such as Ralph Erskine and Vernon Gracie in the Byker area of Newcastle and Rod Hackney in Mackelsfield have made careers in community design and landscape architecture, as have Randolph Hester (1975), Robin Moore, and Ronald Shiffman (J. Stern, 1989) in the United States.

The design process advocated by people such as Rod Hackney relies heavily on a subjective analysis and designing process with the community. This is done either through the participation of a sample of its population or through its leaders articulating the community's needs. Design solutions emerge through the cooperation and/or response to the designs of professionals who provide the technical expertise. The professionals also bring the attention of the community to possibilities that they had not considered (Knevitt, 1985).

Community design is a normative approach to design, limited in the degree to which it has been and is being applied, as it falls outside the general self-image of what an architect is (Saint, 1983). The British architects who have made their careers within this approach have received the measure of prestige, or notority, usually reserved for those architects operating within the mode of buildings as architecture or architecture as art. Those working in the United States have relied much more on environment–behavior research and have been major contributors to it. They have not, however, received the attention of their European counterparts who have relied much more on intuition and are thus, procedurally, closer to the mainstream of the design fields.

In a recent survey, 74 percent of architects in the USA said that they believe that user input produces better design (Dixon, 1988). It is not, however, clear from the report what the respondents understood by "user input." Almost all architects, however, believe that one of the roles of the designer is to bring to people's attention ways of doing things which might be beyond their experience. The basic ideological position of community designers is that direct participation by users of the environment is better than relying on environment–behavior theories. Yet a wide range of positive theories derived from the work of environment–behavior research has affected the way community architects address their clients. This range includes work on human privacy needs and territorial behavior, and research on the needs of people at various stages in life cycle and their desired environments.

The Emergence of Landscape Architecture as Environmental Design and as Art

With the defaulting, from physical environmental design concerns, of the fields of architecture and city and regional planning in the 1960s and 1970s, landscape architecture has been thrust into the forefront of the environmental design disciplines as the public has become more aware of the depletion of resources and the pollution of air and water. The goal became to design for regional and local ecological fitness based on the intrinsic suitability of land (McHarg, 1969). There was also an underlying aesthetic philosophy derived from the Anglo-American tradition of the English landscape garden (see Figure 15).

Figure 15. The English landscape tradition.

This broad environmental interest revives a concern the fields had in the early part of the century, when there was a major concern with natural landscapes and landscape systems as mechanisms to achieve a more salubrious environment. The systematic analysis based on scientific knowledge of natural systems established a mode of thinking that has more recently included a concern for the social and psychological aspects of human life (see Zube, 1986a). This work has been extended into urban design with the advocacies of Hough (1984) and Sprin (1984).

The design of open spaces within the city during the last hundred years has been very much influenced by contemporary views of human beings. Thus, we have pleasure parks (1850–1900), parks as instruments of social reform (1900–1930), parks as active recreational facilities (1930–1965), and now as open-space systems serving both ecological and more passive recreational purposes (Cranz, 1982). The changes in normative orientation have been accompanied by the move of educational programs for landscape architects from schools of agriculture to schools of design (Zube, 1986b). Yet the search for an explicit normative theory of the communication of meaning through landscape design was largely given up until very recently. The search for a normative theory beyond Olmstedian socialism and the English Picturesque is now in swing (Howett, 1987).

The leadership of this search has come from architect-urban designers such as Aldo Rossi, Mattheus Ungers, and the Krier brothers, Leon and Rob. Their goal has been to return to a syntactic focus in designing to make the environment readable. They seek designs that give continuity to people's identification with place. To do this many designers are returning to design based on the adjustment of building, urban, and landscape types rather than a program (C. Rowe, 1983). They, however, think of program in a limited modernist sense, and their aesthetic theories differ. Rossi is historical, Rob Krier neoclassical, Leon romantic-Marxist, Ungers rationalist (Jencks, 1986). Colin Rowe perceives the urban environment as a collage of buildings and places reflecting different architectural attitudes (Rowe & Kotter, 1979).

Landscape architecture has also been caught up in the current art movements (see Figure 16). Deconstructivist thought in architecture has had an impact on landscape design. There is the work of Martha Schwartz, who is seeking to deconstruct convention, and Bernard Tschumi, who seeks to "dismantle meaning." Tschumi's techniques of superimposition, montage, and contiguity are derived from cinema (Johnson & Wigley, 1988; see also Cornwell, 1989).

FIGURE 16. Postmodernism in landscape architecture, Harlequin Plaza, Denver, George Hargreaves, SWA Group, principle designer. Reproduced by permission.

THE MAINSTREAM OF CURRENT ARCHITECTURAL THOUGHT: MODERNISM

Despite the attraction of postmodernism and deconstructivism as normative aesthetic alternatives to the failings of modernism, most architectural thinking remains rooted in the modernist attitudes toward theory. The primary shift away from modernism has been the acceptance of architecture, in academic and avant-garde circles, at least, as almost entirely a high-art form. This view is the core of what is perceived to be current normative thinking. Yet much of what is being built today continues to be within the modernist philosophy. This observation is particularly true of corporate architecture. The irony is that an architecture developed for socialist Europe has found its home in capitalist America.

The guiding slogan is "form follows function" with function still being narrowly defined (Belgasem, 1975). Perhaps in interior design there is a more explicit and self-conscious attention being paid to the actual behaviors a building is to house, and to the symbolic function of architecture. The architect's role is changing as the requirements that a building is to fulfill become more complex. It is an increasingly reduced role. Architects want to hang on to the role of artist but give up other responsibilities. This attitude is particularly true within the discipline of architecture. The professional world marches to a different drummer.

One of the most noticeable changes in the practice of architecture— in architecture as a profession—has been the emergence of a more systematic approach to architectural programming (or brief designing). This development is exemplified by the proliferation of books aimed at professional practice (e.g., Palmer, 1981; Preiser, 1978) and articles on environment–behavior research methods that are applicable to architectural programming and/or postoccupancy evaluation (see Sanoff, 1989; Wener, 1989).

Some work done in environment-behavior research has seeped into almost all architectural practice. This transfer of knowledge has occurred (1) through the new generation of architects having taken relevant courses during their formal education, (2) through guidelines used in professional practice, and/or (3) through the reading of works such as the treatises on the pattern language by Christopher Alexander and his colleagues (Alexander *et al.*, 1977; Alexander 1984). The work on imagability of cities of Kevin Lynch (1960) and the work on defensible space by Oscar Newman (1979) are readily turned into noncontroversial design principles by architects. Consequently, both are widely known and read by architects.

The impact of environment–behavior research on architectural design also depends on the type of building being designed. Few architects doing large-scale housing will design without references to the research literature on housing, particularly that in guideline form (e.g., Cooper, Marcus, & Sarkissian [1986] on housing and Regnier & Pynoos [1987] on housing for the elderly). When it comes to the more prestigious types of buildings most architects favor doing—museums, art galleries, office buildings—the reliance is more likely to be on historic prototypes than on recent environment–behavior research for the positive theoretical basis for design decisions.

THE MODERNISTS ALMOST GOT IT RIGHT

There is a basic conclusion to draw from this review of normative design theories during this century. The early modernists had a comprehensive view of design. Yet their work had many failings that resulted from (1) the weakness of their positive theoretical basis and (2) their attitudes toward their user clients, which was, at best, highly patronizing. The normative models used by the modernists, of what are good people, good behavior, and a good environment, were highly egocentric. The normative theory of the modern movement architects was an ideology with an unclear political bias. It was torn between environmental design and art rather than reconciling them. They almost got it right!

The modernists were well intentioned, if arrogant. While it is difficult to think of their intentions as a single generally accepted agenda, many of them recognized the importance of considering architecture and the allied design professions not only as art but as environmental design based on human needs. Architects like Mies van der Rohe had a social and technological concern, especially in housing design, even though they are perceived today often as strictly formalist architects (see Scott Brown, 1976). Yet the modernists did not form a consistent approach to design. The concern for the living environment of people that Mies demonstrated in the *Weissenhofseidlung* in Stuttgart is not reflected in a concern for student life in the design of the Illinois Institute of Technology campus in Chicago.

The writings of the modernists are replete with models of human needs. One of the best known works is that of Hannes Meyer (see Wingler, 1969). His model focused on the needs paramount in the minds of designers and policymakers having to deal with major housing. Thus, it focuses on the need for shelter, and so, on the human being primarily as a biological entity. The need for a model of human needs as a guide to the development of a design ideology was correct. The models the modernists used were limited ones.

I have described my perception of the mistakes in the thinking of the modernists elsewhere (Lang, 1987). They boil down to two major shortcomings: (1) too few variables were taken as constituting "function" and (2) too little concern was paid to individual and group differences—the personality and cultural aspects of use and form preferences that have their manifestation in design. The modernist's models of the environment and of the human being and the interaction of the two were limited. They had a limited positive theoretical base for their work. This can be rectified. We now have the knowledge which was not at the disposal of the modernists. Whether we care to use it depends on our own ideologies.

IMPLICATIONS FOR FUTURE RESEARCH AND APPLICATIONS

In thinking about design theories from an environment and behavior perspective and/or the contribution of environment–behavior research to the design fields, one has to have an ideological position. It cannot be a value-free, so-called objective analysis. Thus, before discussing the contribution of environment–behavior research to design theory, one must articulate a normative theory.

INTEGRATING ART, ENVIRONMENTAL DESIGN, AND SCIENCE: A NORMATIVE THEORY

The two streams of design thought—design as art and design as environmental design—can and should be brought together within what might tentatively be called a neomodernist normative design theory. It might also be called a behavior deterministic theory because it assumes that designing for human behavior, in its multiplicity of complexities, is the purpose of design. The implication is that "form follows function" is a good slogan. We just have to think of function in a much broader way than the modernists did.

The point of departure is to recognize the complexity and multidimensionality of the environment and of "behavior," in its broadest possible meaning, in developing a positive theoretical basis for design. Designers need to understand the relationship between the social and the physical environment, both its natural and artificial components. The two are inextricably linked. Designers need to understand what is potentially there to be perceived and used and what constitutes different people's effective environments or, at least, how to develop this understanding in a particular circumstance. The reason is simple. Changes to one component of the environment affect the others, with possibly both positive and negative consequences.

It is important to think of the environment as a behavior setting as well as a geometry of surfaces and textures. As a behavior setting the environment consists of combinations of milieu and standing patterns of behavior in time (Barker, 1968; Kaminski, 1989). As geometry, the environment consists of surfaces that, pigmented and illuminated, carry associated meanings and may, although it now seems unlikely, *per se,* express emotion. Over time these patterns of behavior and perceptions of the quality of fit between environment and behavior, including aesthetic values, change. The changes may or may not require changes in the layout of the environment to accommodate them. This depends on the affordances of the patterns of the environment and on standards of satisfaction applied to the misfit between environment and behavior.

The primary concern of physical design is with having the layout, or structure, of the physical environment meet human needs in a technologically sound manner within the constraints of the resources available. This may involve changing the existing layout of the environment or preserving it. The specific need or set of needs to be the focus of attention will vary from problem to problem, but they all need to be addressed. As basic needs become fulfilled so the perception of need shifts. A number of models of human needs are available to designers (e.g., Leighton, 1959; Maslow, 1954). Better ones (i.e., ones that explain more about human life and purpose) will almost certainly be developed in the future, but they have been slow in coming. Maslow's model, despite its shortcomings, has a comprehensive nature and hierarchical organization that enables us to consider the concept of function of the artificial environment in a broader way than the modernists did. The design concerns at each level of needs, as identified by Maslow, are summarized in Table 1.

Using such a model as the basis for asking questions about the purposes of a design or proposed design implies that aesthetics is a function—a need to be fulfilled. An artist's need as a contributor to society is important to consider but so are the needs of users and sponsors of projects. As values will differ, and designers will always work with imperfect knowledge, the design process must be considered to be an argumentative process.

The overall objective of improving the quality of the environment for people is to create settings and/or new aesthetic displays for these human needs to be met. In dealing with other species other needs have to be met, but ultimately these serve human needs too. There are various design and planning fields involved in the creation of settings. In a capitalist, democratic society they are established by a combination of governmental activity and the activity of those individuals and groups who hold the financial wealth in society. The range of settings provided depends on the perceptions of those making decisions and the resources

TABLE 1. Human Needs and Design Concerns

Need	Design concerns/sociophysical mechanisms
Survival	Shelter; access to services
Safety	Access to services; privacy; territorial control; orientation in society, time, and space
Belonging	Access to services and communal settings; symbolic aesthetics
Esteem	Access to services; control; personalization; symbolic aesthetics
Self-actualization	Choice; control; access to developmental opportunities
Cognitive/aesthetic	Access to developmental opportunities; formal aesthetics; art for art's sake

available. Even the poorest person in such a society, however, has the ability to adjust some aspect of his or her environment.

Some new behavior settings require the physical layout of the environment to be changed. Thus, the process of design can be seen as a process of adapting the layout of the world to changing manifestations of human needs within a terrestrial, social, and cultural context. A host of fields deal with the design of the social and economic environment. Planned intervention into the normal operation of the market and legal system takes place at a variety of levels. The various design fields operate within the norms of the market, but they also attempt to change national policy. Sometimes this is in terms of their own self-interests and sometimes on behalf of what they see would lead to a better society. As they are now constituted, the design fields have expertise in various aspects of changing the physical enviornment. These skills and the knowledge base required to predict the outcomes of design will have to be continually developed as society changes. What is important is for designers to be able to predict accurately the outcome of their work.

In any society there are likely to be differences in opinion about what changes in the design, or layout, of the world should be made. There are usually differences in opinion over what the problem is. Thus, designers almost always end up advocating for one person's ends over another's, depending on their own ethical positions and their willingness to depart from their professed positions.

Every designer, or set of designers, addressing a problem has a way of working through the design process. Implicit, if not explicit, in his or her work is a normative procedural theory. A set of techniques is used to identify what the problem is, to generate possible solutions to that problem, to evaluate how these solutions will work in the future, and to decide on one course of action, to implement it, and to evaluate the results. The set of techniques used depends on the knowledge designers have of the range of techniques available and their knowledge about how the environment works and the interaction of the proposed designs and their contexts. Thus, the fundamental concern of designers, I argue, has to be with environmental design. Yet the artist and the artistic endeavor is an important part of society. They enrich life immeasurably.

Every creation of designers can be viewed as an object—a display—or as an environment. These are two modes of looking at the world (Gibson, 1979). We do create displays for other people sometimes in a concerted action at a city scale (Olson, 1986) and sometimes in the creation of a miniature painting. There will almost always be a component of pure fine art in designing the physical environment—a place where designers can present their own personal expression through design.

The political question that arises is: How far should the designer be expected to go? Certainly not to the extent that the artistic expression is detrimental to the lives of people. Ideally it should support them.

How far should designers go in imposing and fighting for their own visions of a good society? We get so used to accepting standard types of environments. The design process is often one of simply adapting existing building and urban types and constructing them in new places. We no longer think of the consequences of this. National policy and design philosophy at the regional and urban scale has led to all kinds of segregations—activities, age groups, and the able from the disabled, for example. Should this continue?

Design is an argumentative process. It is my position that these arguments should be based on knowledge of how the world works, how the natural interacts with the artificial, and how the artificial interacts with the various species that inhabit the world. This is by no means a unique position. It is professed by both designers and environment–behavior researchers. Yet, in practice, the position on issues other than the technology of design is often different. This must continue to change if the roles of designers as professionals in society are to remain a service to society.

If one looks at the environment and the design fields in this way, it is clear that we need to know much more than we do. Environment–behavior research is essential if we designers are to fulfill our potential roles in society. We researchers need to be concerned with the disciplines of design as a whole, not with either narrow professional definitions or claims over the market. Contributions to the disciplines will enhance professional performance, if not in the short run, then certainly in the long run. The contributions may indeed lead to a complete overhaul of professional definitions and possibly to less competition and less fragmentation of efforts to improve the habitat of people.

Our procedural models need to be rethought. First of all, one needs to recognize that designing is value laden. The designer's role is greater than that which it is considered to be by community designers. Designers are more than midwives. While designers must listen to people, they (along with other experts) also need to be able to bring the attention of clients—users and sponsors—to different social organizations and aesthetic expressions and thus to different design possibilities. Clearer models of the design process and new techniques of analysis, synthesis, prediction, and evaluation will be needed if designers are to maximize their potential contributions to society. As artists and environmental designers, the design professionals have a special obligation to society. They are holders of a unique expertise that they have a moral obligation to develop. Environment–behavior research, both substantive and procedural, is central to this endeavor.

THE ROLE OF ENVIRONMENT–BEHAVIOR RESEARCH: BUILDING THE DESIGN DISCIPLINES

The role of environment–behavior research is, indeed, fundamental to the building of the design disciplines, if any image of the designer other than that of pure fine artist is accepted. If the attitude toward design expressed in this chapter is accepted, then a scientific or quasi-scientific basis is necessary for it to be done well. This basis is necessary because there are limitations to human common sense and to an individual designer's experience with the world. It is necessary because it proposes that designers shift the design process away from one based on design by habit, using self-referent information, to one based on an explicit body of positive theory—substantive and procedural. This will demand more of a designer's creative thinking.

The contribution of environment–behavior research is, however, even broader than that suggested in the above paragraph. It will be to challenge the very nature of theory for the design fields. Borrowing liberally from the work of others, I have proposed, here (see Figure 1) and elsewhere (Lang, 1980, 1987), a quadrapartite model of theory for the design fields. This model is, however, a tentative one and is already being questioned (e.g., Vernez-Moudon, 1988). Its explicitness opens it to critical examination and thus to rejection when a better model is created. What is needed is a metaphysics of the design fields. This will be the overriding challenge for environment–behavior and design researchers.

A second contribution will be the enhancement of the positive substantive basis for design—the understanding of how patterns of the environment interact with patterns of human spatial and emotional behavior. This contribution will occur at two levels. The first is at the broad scale of the understanding of human nature and needs and their implications for design: the discussion of the model of the human being used as a basis for design is fundamental to design theory (see Ellis & Cuff, 1989). The second is in understanding how patterns of the environment meet specific requirements. The goal of the first level is to enhance the quality of the argument over what the design objectives should be in a particular circumstance, and the goal of the second level is to understand the design principles required to meet those objectives. All design criticism boils down to questions/arguments over (1) human goals, (2) design objectives, and (3) the principles for achieving both. Distinguishing among them helps focus the attention of discussions about a design.

The third major contribution to design is and will be a methodological one. It is a twofold contribution: methods of problem solving and methods of research (see Zeisel, 1981). Problem-solving methods are those involving the analysis, design, and evaluation of the environment and proposed patterns of the environment. Research methods are those for obtaining information about the world and explaining why the world is the way it is. The methods overlap; some are the same. Conceptually, the latter can be scientific, although this is not easy because of the complexity—the multiplicity of variables and the difficulty of quantifying them—of the issues of concern to designers.

The goal of problem-solving methods is to help the designer do a job, to build well. The goal of basic research is to build theory that informs design. Theory is thus an intermediary between research and practice (Lang, 1988a). Theory is and will be informed by learning from practice. In the future it will be informed even more by environment–behavior research.

In the past, environment–behavior research has informed theory on an ad hoc basis. It will continue to do so. It will occur more explicitly in the future as models of theory are developed based on their utility for practice. In the meantime, the contribution will continue to be through the development of guidelines for design. This should not, however, be the sole goal.

Guidelines have inherent advantages for designers, principally their ease of use—the designer does not have to understand much about the phenomena of concern. Guidelines also have inherent disadvantages, the most important one being the tendency to see the situation being faced as the same as that for which the guidelines were developed—the availability fallacy. Christopher Alexander recognizes this about his own work, which is presented essentially in guideline form (Alexander *et al.*, 1977). Thus, say, in designing for the elderly, a designer can follow the guidelines, or design directives, proposed by Regnier and Pynoos (1987) and their colleagues. These guidelines are based on painstaking research, but the designer does not have to understand much about the elderly to use them. Being able to put these guidelines into a theoretical framework will enhance the problem-solving abilities of the designer. That is the ultimate goal of environment–behavior research.

The desire to do research that is professionally applicable is laudable, but it should not obscure the broader goal. The need is to develop the universe of potentially useful knowledge for designers—the whole domain of environment–behavior research—and not simply what seems to be immediately useful in professional practice. The contribution is to the development of positive theory for design and an understanding of normative theories and their implications.

REFERENCES

Alberti, L. B. (1955). *Ten books on architecture* (J. Leoni, Trans., 1726). London: Tiranti. (Original work published 1755)

Alexander, C. (1984). *The search for a new paradigm in architecture.* Boston: Oriel.

Alexander, C., Ishikawa, S., & Silverstein, M. (1977). *A pattern language.* New York: Oxford University Press.

Altman, I. (1975). *The environment and social behavior.* Monterey, CA: Brooks/Cole.

Anderson, S. (1988). *Some themes for a symposium on Ph.D. education in architecture.* Unpublished manuscript, Department of Architecture, Massachusetts Institute of Technology, Cambridge.

Banham, R. (1960). *Theory and design in the first machine age.* New York: Praeger.

Barker, R. (1968). *Ecological psychology: Concepts and methods.* Stanford, CA: Stanford University Press.

Belgasem, R. (1987). *Human needs and building evaluation: An approach to understanding architectural criticism.* Unpublished doctoral dissertation, University of Pennsylvania, Philadelphia.

Benevolo, L. (1971). *The origins of modern town planning* (J. Landry, Trans.). Cambridge, MA: MIT Press.

Blake, P. (1974). *Form follows fiasco.* Boston: Atlantic-Little.

Broadbent, G. (1973). *Design in architecture: Architecture and the human sciences.* New York: Wiley.

Brolin, B. (1976). *The failure of modern architecture.* New York: Van Nostrand Reinhold.

Collins, P. (1965). *Changing ideals in modern architecture 1750–1950.* Kingston and Montreal: McGill-Queen's University Press.

Conrad, U. (1970). *Programs and manifestoes on twentieth-century architecture.* Cambridge, MA: MIT Press.

Cooley, C. H. (1909). *Social organization: a study of the larger mind.* New York: Scribner.

Cooper Marcus, C., & Sarkisssian, W. (1986). *Housing as if people mattered.* Berkeley: University of California Press.

Cornwell, R. (1989) MoMA's ego builder. *New Art Examiner, 16*(6), 36–41.

Cullen, G. (1962). *Townscape.* London: Architectural Press.

Cranz, G. (1982). *The politics of park design.* Cambridge, MA: MIT Press.

Dixon, J. M. (1988). P/A reader poll design preferences. *Progressive Architecture, 69*(10), 15–17.

Eaton, L. (1969). *Two Chicago architects and their clients: Frank Lloyd Wright and Norman Van Doren Shaw.* Cambridge, MA: MIT Press.

Eisenmann. P. (1982). *House X.* New York: Rizzoli.

Ellis, R., & Cuff, D. (1989). *Architect's people.* Berkeley: University of California Press.

Eslami, M. (1985). *Architecture as discourse: The idea of method in modern architecture.* Unpublished doctoral dissertation, University of Pennsylvania, Philadelphia.

Festinger, L. (1957). *A Theory of cognitive dissonance.* Evanston, IL: Row, Petersen.

Foster, H. (1985). *Recoding: Art, spectacle, cultural politics.* Port Townsend, WA: Bay Press.

Gibson, J. J. (1950). *Perception of the visual world.* Boston: Houghton Mifflin.

Gibson, J. J. (1966). *The senses considered as perceptual systems.* Boston: Houghton Mifflin.

Gibson, J. J. (1979). *An ecological approach to visual perception.* Boston: Houghton Mifflin.

Gideon, S. (1941). *Space, time and architecture: The growth of a new tradition.* London: Oxford University Press.

Gottschalk, S. S. (1975). *Communities and alternatives: An exploration of the limits of planning.* Cambridge, MA: Schenkman.

Gray, C. (1953). *Cubist aesthetic theories.* Baltimore, MD: Johns Hopkins University Press.

Groat, L., & Canter, D. (1979). A study of meaning: Does Post Modernism communicate? *Progressive Architecture, 60*(12), 84–87.

Hester, R. (1975). *Neighborhood space.* New York: Van Nostrand Reinhold.

Holden, A. (1988, October 30). The crusader prince. *Sunday Times,* C1–C2.

Hough, M. (1984). *City form and natural process.* New York: Van Nostrand Reinhold.

Howard, E. (1902). *Garden cities of tomorrow.* London: Sonnenschein.

Howett, C. (1987). Systems, signs, sensibilities. *Landscape Journal, 6*(1), 1–11.

Hubbard, H. V., & Kimball, T. (1917). *Landscape architecture.* New York: MacMillan.

Ittelson, W. H. (1973). Environmental perception and contemporary perceptual theory. In W. H. Ittelson (Ed.), *Environment and cognition* (pp. 1–19). New York: Seminar Press.

Jacobs, J. (1961). *The death and life of great American cities.* New York: Random House.

James, W. (1890). *The principles of psychology.* New York: Holt.

Jencks, C. (1986). *What is post-modernism?* New York: St. Martin's Press.

Jencks, C. (1988). De construction: The pleasure of abuse. *Architectual Design, 58*(3/4).

Johnson, P., & Wigley, M. (1988). *Deconstructivist architecture.* New York: Museum of Modern Art.

Jones, J. C., & Thornley, D. G. (Eds.). (1963). *Conference on design methods.* New York: MacMillan.

Kaminski, G. (1989). The relevance of ecologically oriented conceptualizations to theory building in environment and behavior research. In E. H. Zube & G. T. Moore (Eds.), *Advances in environment, behavior, and design* (Vol. 2, pp. 3–36). New York: Plenum.

Khan-Mahomedov, S. O. (1971). N. A. Ladovsky, 1881–1941. In O. A. Shvidovsky (Ed.), *Building in the USSR* (pp. 72–76). New York: Praeger.

Klotz, H. (1989). *The history of postmodern architecture* (R. Donnell, Trans.). Cambridge, MA: MIT Press.

Knevitt, C. (1985). *Space on earth: Architecture, people, and buildings.* London: Thames Methuen.

Koberg, D., & Bagnall, J. (1974). *The universal traveler.* Los Altos, CA: Kaufmann.

Lang, J. (1980). The nature of theory for architecture and urban design. *Urban Design International, 1*(2), 41.

Lang, J. (1987). *Creating architectual theory: The role of the behavioral sciences in environmental design.* New York: Van Nostrand Reinhold.

Lang, J. (1988a). Theory as an intermediary between research and practice in architecture. *Design Methods and Theories, 22*(5). 15–24.

Lang, J. (1988b). Understanding normative theories of architecture. *Environment and Behavior, 20*(5), 601–632.

Le Corbusier (1967). *The radiant city.* New York: Orion Press. (Original work published 1934)

Le Corbusier (1968). *The modular* (P. de Francia & A. Bostock, Trans.). Cambridge, MA: MIT Press. (Original work published 1951)

Le Corbusier (1970). *Towards a new architecture* (F. Etchelles, Trans.). New York: Praeger. (Original work published 1923)

Le Corbusier (1973). *The Athens charter* (A. Bardley, Trans.). New York: Grossman.

Leighton, A. H. (1959). *My name is Legion: Foundations for a theory of man in relation to culture.* New York: Basic Books.

Lynch, K. (1960). *The image of the city.* Cambridge, MA: MIT Press.

Maslow, A. (1954). *Motivation and personality.* New York: Harper & Row.

McHarg, I. (1969). *Design with nature.* New York: Doubleday.

Michelson, W. (1976). *Man and his urban environment: A sociological approach.* Reading, MA: Addison-Wesley.

Moore, G. T. (Ed.). (1970). *Emerging methods in environmental design and planning.* Cambridge, MA: MIT Press.

Moore, G. T. (1988). *Toward a conceptualization of environment-behavior and design theories of the middle range.* Paper presented at the biennial conference of the International Association for the Study of People and their Physical Surroundings, Delft, the Netherlands.

Newman, O. (1979). *Community of interest.* New York: Anchor.

New York Regional Survey. (1979). New York: Regional Plan Association.

Norris, C., & Benjamin, A. (1989). *What is deconstruction?* London: Academy Editions.

Olson, D. (1986). *The city as a work of art.* New Haven, CT: Yale University Press.

Overy, P. (1969). *Kandinsky: The language of the eye.* New York: Praeger.

Palmer, M. A. (1981). *The architect's guide to facility programming.* Washington, DC: AIA and Architectural Record Books.

Peña, W. (1977). *Problem seeking.* Boston, MA: Cahners.

Preiser, W. (Ed.). (1978). *Facility programming: Method and applications.* New York: Van Nostrand Reinhold.

Rapoport, A. (1977). *Human aspects of urban form.* New York: Pergamon Press.

Rasmussen, S. E. (1959). *Experiencing architecture.* Cambridge, MA: MIT Press.

Regnier, V., & Pynoos, J. (1987). *Housing the aged: Design directives and policy considerations.* New York: Elsevier.

Rowe, C. (1983). Program vs. paradigm. *Cornell Journal of Architecture, 2,* 8–19.

Rowe, C., & Koetter, F. (1979). *College city.* Cambridge, MA: MIT Press.

Rowe, P. (1987). *Design thinking.* Cambridge, MA: MIT Press.

Saint, A. (1983). *The image of the architect.* New Haven, CT: Yale University Press.

Sanoff, H. (1989). Facility programming. In E. H. Zube & G. T. Moore (Eds.) *Advances in environment, behavior, and design* (Vol. 2, pp. 239–286). New York: Plenum.

Santayana, G. (1955). *The sense of beauty.* New York: Dover. (Original work published 1896)

Scott Brown, D. (1976). On formalism and social concern: A discourse for social planners and radical chic architects. *Oppositions, 5,* 99–112.

Senkevitch, A. (1974). *Trends in Soviet architectural thought 1917–1937.* Unpublished doctoral dissertation, Cornell University, Ithaca, NY.

Smithson, A. (1968). *Team 10 primer.* Cambridge, MA: MIT Press.

Smithson, A., & Smithson, P. (1967). *Urban structuring.* New York: Reinhold/Studio Vista.

Spirn, A. (1984). *The granite garden.* New York: Basic Books.

Stein, C. (1951). *Towards new towns for America.* New York: Van Nostrand Reinhold.

Stern, J. (1989). Pratt to the rescue: Advocacy planning is alive and well in Brooklyn. *Planning, 55*(5), 26–28.

Stern, R. A. M. (1988). *Modern classicism.* New York: Rizzoli.

Tauber, M. L. (1973). *Abstract art and visual pereption.* Paper presented at the 61st Annual Meeting of the College Art Association, New York.

Thiel, P. (1961). A sequence-experience notation for architectural and urban spaces. *Town Planning Review, 32,* 33–52.

Venturi, R. (1966). *Complexity and contradiction in architecture.* New York: Museum of Modern Art.

Vernez-Moudon, A. (1988). *Normative/substantive and etic/emic dilemma in design education.* Unpublished manuscript, College of Architecture, University of Washington, Seattle.

Vitruvius (1960). *The ten books on architecture* (H. H. Morgan, Trans., 1910). New York: Dover.

Wener, R. (1989). Advances in evaluations of the built enviornment. In E. H. Zube & G. T. Moore (Eds.), *Advances in environment, behavior, and design* (Vol. 2, pp. 287–313). New York: Plenum.

White, M., & White, L. (Eds.). (1964). *The intellectural versus the city from Thomas Jefferson to Frank Lloyd Wright.* New York: New American Library.

Wines, J. (1985). *De-architecture.* New York: Rizzoli.

Wingler, H. (1969). *The Bauhaus* (W. Jabs & B. Gilbert, Trans.; J. Stein, Ed.). Cambridge, MA: MIT Press.

Wolfe, T. (1981). *From Bauhaus to our house.* New York: Farrar, Straus & Giroux.

Wright, F. L. (1958). *The living city.* New York: Horizon Press.

Wright, F. L. (1960). Prairie architecture. In E. Kaufman & B. Raeburn (Eds.), *Frank Lloyd Wright: Writings and buildings.* New York: Horizon Press.

Young, M., & Wilmott, P. (1957). *Family and kinship in East London.* London: Routledge and Kegan Paul.

Zeisel, J. (1981). *Inquiry by design: Tools for environmental-behavior research.* Monterey, CA: Brooks/Cole.

Zube, E. H. (1986a). The advance of ecology. *Landscape Architecture, 76*(2), 58–67.

Zube, E. H. (1986b). Landscape planning education: Retrospect and prospect. *Landscape and Urban Planning, 13*(56), 367–378.

3

Relationships between Research and Design
A COMMENTARY ON THEORIES

MARTIN S. SYMES

The emphasis in this volume on relationships between environment–behavior research and design studies represents a new direction for the series. In the first volume, Franck (1987) hoped that the introduction of new theoretical perspectives would strengthen the links between psychology and design. Readers have had to wait for the current volume to be exposed to extensive discussion of the intellectual framework which will be required. It is undoubtedly a healthy sign that the authors addressing this issue argue strongly for the possibility of developing a unified theoretical stance. Their arguments are, however, mainly pragmatic. Equally strong arguments can be mounted for suggesting that integration at this level is unlikely to occur. In addition, there are a number of alternative philosophical positions available for environment-behavior research and no clear rationale for selecting only one.

This commentary aims to review the arguments for and against creating an integrated theoretical framework for research studies in the two fields of environment–behavior studies and design. It begins with a critical discussion of the cases for creating such a framework put forward by Groat and Després and by Lang. It then outlines a contrary case—

Martin S. Symes • Department of Architecture, University of Manchester, Manchester M13 9PL, United Kingdom.

that against attempting to create a single theoretical framework for this field.

The case for integration seems to rest on three propositions, each of which can be questioned. The first is that environment–behavior and design theories have common roots. It is argued that present differences between them are temporary and that their resolution should be anticipated. Against this it can be argued that the trend has been for these theoretical positions to diverge and that convergence is unlikely. In addition, it is difficult to identify any published research that has attempted this integration. The second proposition is that the differences between environment–behavior and design theories are not fundamental, but rather in the nature of misunderstandings. Two of these are singled out by the authors: the conflation of disciplinary and professional knowledge, and the loose interpretation of the term "theory." The solution to this problem is not, however, to rule out some knowledge and certain interpretations—it is to recognize that these are different types of knowledge and different types of theory, all of which need clearer differentiation.

The third proposition in the case for an integration of perspectives is that this is more likely to lead to the development of useful knowledge. But against this it may be demonstrated that much useful knowledge has already been produced within the separate fields as they are currently defined. Thus, all three main propositions in the authors' cases have been given a pragmatic basis but may be refuted.

The alternative case, that against attempting to create a single integrated framework, is constructed at two levels. At one level, the pragmatic, important problems concerning the constitution of the total field and the definition of its boundaries are as yet unsolved. Possibly they should remain open questions; if so, a single integrative theory would be inappropriate. At a second level, the paradigmatic, alternative assumptions can be made concerning the nature and purpose of scientific inquiry. These may not all be appropriate for the whole field. In the related area of planning studies, the "liberal position" is thought to omit important insights through its analytical approach to the subject matter. "Alternative positions" seek links and connections. Some of these may be just as relevant in environment–behavior–design research.

THE CASE FOR INTEGRATION

The emergence of a new field of inquiry is an event of some considerable interest for those working in the subject area concerned. Some are interested in areas that are presently parts of other fields. Some who work on related topics may need to take account of the new depth of knowledge generated. The emergence of a new field of inquiry is thus a social as well as intellectual event. It is likely to be contested, before acceptance, on both social and intellectual grounds.

Research on environment, behavior, and design was first given some degree of recognition in the late 1960's. It is still not entirely clear that all parties accept the existence of the field of inquiry or of the professional group that goes with it as permanent features of academic life. The authors of both previous chapters present the case for the coherence of work in this field in normative terms—as work that should be possible rather than as work that has been achieved.

History

Both chapters support the case for theoretical integration with a historical review of approaches to the study of human behavioral requirements by designers. Groat and Després trace the origins of this interest to the time of the Renaissance. They give a brief account of how architectural theory, defined as rhetoric and as principles, has related to the development of a scientific epistemology and of the social sciences. Lang's treatment is restricted to the last hundred years or so but deals in more depth with the variety of views that have emerged.

The argument is that social science and contemporary architectural theory have certain roots in common and share similar goals and objectives. Unfortunately, the case for still considering them as an integrated field is not strong. Lang's demonstrations of the differences between those who argue as artists and those who argue as environmental designers suggest divergence rather than convergence as the historical trend.

Groat and Després make a sophisticated distinction between philosophical positions taken by deconstructivists and environmental researchers, suggesting that the disjuncture is temporary. This seems unlikely. The experience in many other disciplines has surely been that of increasing differentiation of viewpoints. Philosophy itself is often thought to have roots in only two traditions, the Platonic and the Aristotelian, but numerous new approaches have grown out of them, each eventually finding its place in a galaxy of positions now available to the modern thinker. Design-related research can surely follow a similar trajectory.

In addition, it should be pointed out that the literature of the field does not seem to support the view that there is essentially a single tradition that is ready to reconverge.

The case would be stronger if there were already research publications in the environment–behavior and design fields with such comprehensive interests as Swenarton's (1989) report on the influence of Ruskin's moral values on early twentieth-century architects or Banham's (1960) documentation of the enduring importance of symbolism in machine-age design.

Misunderstandings

Both chapters suggest that misunderstandings have arisen from the juxtaposition of two research traditions and that if these can be removed, communication between those working in architectural theory and environment–behavior research will be improved.

The conflation of discipline and profession.   The basic proposition made by both chapters is that only some aspects of the work reported in the literature refer to the development of an academic discipline. The remainder is seen as practical information for the profession and part of a separate discussion. The perception of beauty would fall into the former category. The marketing of services would fall into the latter. This is a simplistic argument. The misunderstanding stems only from the inability of students of aesthetics, for example, to categorize issues in marketing. But it is incorrect to suggest that there is no practical application for knowledge about perceptions. Rather than implying that professional knowledge could be omitted from the debate, we should perhaps call for clarity about levels of abstraction between the professional and the academic. Such levels can surely be found in all the subject areas of environment–behavior–design studies.

Saint (1983) shows how, at different periods in history, architects have laid emphasis on the entrepreneurial aspects of their trade and on its artistic ambitions. A similar observation should be made about the changing emphasis of research work in universities. All those involved have to make their living from thinking about their specialism, sometimes this need figures more strongly, sometimes less.

Alternative terminologies.   The discussions in both chapters reveal that the literature lends itself to considerable confusion. Designers tend to discuss "theories" of what should be done, while scientists refer to "theories" of what may be the case. Happily, there is some consensus over the means of achieving an appropriate clarification. In particular,

Lang makes two useful distinctions, that between substantive and procedural theory and that between positive and normative theory. It certainly will be of great value in the future growth of this field for theorists to elaborate in more detail the four types generated by Lang's classification. Distinct contributions can surely be made to the development of each type of theory.

USEFUL KNOWLEDGE

Lang calls for more procedural knowledge, describing what designers and design researchers actually do, or what they could or should do. Research into procedures will be useful to practitioners as well as to those who commission or evaluate their work.

The authors of both chapters claim that substantive research can and should be further pursued. Lang asserts that the "modernists almost got it right." The field now offers knowledge that will allow practitioners to avoid the faults of earlier decades. He points to better models of environments, of people, and of their interaction, more attention to individual and group differences, and a less limited concept of building function. Groat and Després also make specific proposals for the elaboration of physical variables, such as type, morphology, composition, and style, in hypothesis generation, empirical research, and theory building.

There is no question that work which uses these models and these variables has been initiated and is beginning to bring interesting results. The argument made in these two chapters, that such research should be undertaken with an integrated single discipline and not under the aegis of different disciplines in communication with each other, does not necessarily follow. Some examples will make the point. Schon's (1983) work on design procedures has been pursued within the discipline of sociology. Saint's (1987) work on the social history of a particular building type has been undertaken within the discipline of history. They seem not to have suffered from academic isolation. But Hillier and Hanson's (1984) provocative studies on urban spatial organization and Stiny's (1985) work on the generation of architectural form could arguably benefit from more exposure to criticism from other disciplines. Work by Oliver (1969) on vernacular architecture could perhaps be deepened with cross-disciplinary approaches of the type so productively pursued by Rapoport (1969).

Useful knowledge may well come from more integrated work in this heretofore multidisciplinary area, but it would be unwise to maintain that no more insights are likely to arise from the constituent disciplines pursuing their own interests in their own way. Both approaches have something to offer.

THE CASE FOR DIVERSITY

The case for diversity is partly a social one. Research is currently being undertaken in a variety of contexts. It may be as difficult to extract the people concerned from their context so that they may work together as it would be to facilitate communication among them by scientific unification. The case for diversity is also an intellectual one, and it is this which will be elaborated here.

UNPREDICTABLE SYNTHESIS

The first point in the case for diversity is that the best intellectual basis for an integration of the field may prove difficult to predict.

As Kuhn (1962) showed, new theoretical developments, revolutionary science, usually arise in response to intellectual or practical uncertainties. There are a number of these in environment-behavior-design research at the present time, but their origin and source is rarely predictable. It would be unwise to assume that the innovations required to help improve the quality of the environment for its users will come only from the disciplines engaged in current research. For example, March (1976) shows that a valuable contribution has recently been made by mathematics. Is this the core discipline of the future? Again, no contribution had come from political science until Schon (1983) began writing about design method. Is this a discipline from which special knowledge will continue to be drawn? The questions are open and should arguably remain unanswered.

The pursuit of a continuum of levels of abstraction in research methodologies, from the most academic to the most practical, is fundamental to this commentary. It contrasts with the purpose of the previous chapters. This appears to be the maintenance of an academic discipline distinct from its associated practical professions. Other authors (e.g., Joroff & Moore, 1984) have suggested that connections not only should but also can be made. There is clearly no consensus here. A debate over the possibility of an integrated field should certainly be held. Whether it can, or should, be settled one way or another is once more an open question.

VARIETY OF PARADIGMS

The second point in the case for diversity is that there are underlying questions about scientific method that have yet to surface fully in this field. The assumptions of rationality and control as a behavioral goal that were adopted a century ago have been more clearly questioned in relation to planning studies than to architectural design. The criticisms apply equally well to all problems in environmental design that involve user requirements. To discuss the form of integration in this field of study without addressing these assumptions may lead to more confusion than it will avoid.

The liberal position. This term is used by Marris (1987) to describe a method that separates a research topic into discrete aspects and then studies them independently. He seems to refer to an analytical philosophical position and a hypothetico-deductive experimental method such as is often associated with the "systems approach" to planning and the "quantitative revolution" in geographical studies. In this approach, planners use the results of physical measurement and quantitative modeling as the basis for policy generation. The possibilities for planning control that follow are characterized as dependent on socially agreed-upon economic goals.

Critics of this approach maintain that basic human interests often cross the boundaries of analytic categories. The liberal method breaks up these interests, gives priority to some aspects, and leads to conflict between policy areas. It may also lead to misunderstanding by those affected. Albrecht and Lim (1986), for example, argue that planners should be committed to more comprehensive social reform and that they have often been known to think and act in such a disjointed, incremental manner that they cannot be understood or provide profound alternatives.

In suggesting that we treat design principles as hypotheses "testable" in principle, the authors of these two chapters still adopt Marris's liberal position. They frame their ideas in a tradition that has been questioned from a political point of view. In so far as Lang and Groat and Després frame their ideas in this analytical tradition, they run the risk that the research they appear to sponsor will be deemed inappropriate.

Critical positions. Marris's answer to this kind of concern is to propose an alternative paradigm in his particular area, community planning. His proposal is that decisions should be locally based. Research should seek links and connections between phenomena. Social science should be concerned with the recovery of meaning. Environmental studies should be constructed around ethnographic methodologies.

Groat and Després, as well as Lang, are, of course, aware of ethnographic methodologies and make special mention of phenomenology. This has gained some popularity in landscape analysis and has had some influence on the rather confused state of architectural theory today. Differential values, intersubjectivity, and the need for reflection on action are issues more readily discussed in the phenomenological perspective than in the liberal position.

The distribution and exercise of power is also a factor in environmental policy and one not easily studied in a scientific way. For studies which move into this area, some methodologists advocate the currently fashionable "critical theory." Frampton (1988) is an architect writing in this vein. He argues that analytic methodologies have failed to produce information of value to designers. They should ally themselves with local interests in each of their projects. They should resist the fragmentation that stems from rationalized knowledge. They should see examples of what he calls "critical practice" as another, more valuable type of knowledge.

With which of these methodologies should appropriate questions for environment–behavior–design research be framed? Surely the community of researchers and practitioners has not yet decided. Thorough and clear-sighted chapters by Groat and Després and by Lang have brought us to the position where we may begin to hold a debate. Let us hope that a future volume in this series will see the questions they raise given a more complete answer.

REFERENCES

Albrecht, J., & Lim, G. C. (1986). A search for alternative planning theory: Use of critical theory. *Journal of Architectural and Planning Research, 3*(2), 117–131.

Banham, R. (1960). *Theory and design in the first machine age.* London: Architectural Press.

Frampton, K. (1988). Place-form and cultural identity. In J. Thackara (Ed.), *Design after modernism* (pp. 51–66). London: Thames & Hudson.

Franck, K. (1987). Phenomenology, positivism, and empiricism as research strategies in environment–behavior research and in design. In E. H. Zube & G. T. Moore (Eds.), *Advances in environment; behavior, and design* (Vol. 1, pp. 59–67). New York: Plenum.

Hillier, B., & Hanson, J. (1984). *The social logic of space.* Cambridge, England: Cambridge University Press.

Joroff, M. L., & Moore, J. A. (1984). Case method teaching about design process management. *Journal of Architectural Education, 38*(1), 14–17.

Kuhn, T. (1962). *The structure of scientific revolutions.* Chicago: University of Chicago Press.

March, L. (Ed.). (1976). *The architecture of form.* Cambridge, England: Cambridge University Press.

Marris, P. (1987). *Meaning and action: Community planning and conceptions of change.* London: Routledge & Kegan Paul.

Oliver, P. (Ed.). (1969). *Shelter and society.* London: Barrie & Rockliff.

Rapoport, A. (1969). *House form and culture.* Englewood Cliffs, NJ: Prentice-Hall.

Saint, A. (1983). *The image of the architect.* New Haven, CT: Yale University Press.

Saint, A. (1987). *Towards a social architecture: The role of school-building in post-war England.* New Haven, CT: Yale University Press.

Schon, D. (1983). The reflective practitioner. New York: Basic Books.

Stiny, G. N. (1985). Computing with form and meaning in architecture. *Journal of Architectural Education, 39*(1), 7–19.

Swenarton, M. (1989). *Artisans and architects: The Ruskinian tradition in architectural thought.* London: Macmillan.

II

ADVANCES IN PLACE RESEARCH

4

Workplace Planning, Design, and Management

FRANKLIN BECKER

As part of a general management theory, Frederick Taylor's *Principles of Scientific Management* (1911) marked a great turning point in our thinking about how the planning, design, and management of the workplace affects individual and organizational performance. With respect to physical design, Taylor did much more than simply suggest how factors such as office layout affect work patterns. He articulated a fundamental philosophy of office planning and design that almost 80 years later is still a powerful force influencing how offices are designed and planned in North America. Its hallmark was efficiency. Furniture and plant layouts that minimized "wasted" movement (and thus wasted time and money) were the goal.

Over the next three-quarters of a century, various other issues besides efficiency have surfaced to guide research and writing about workplace planning and design. For the most part these have coexisted with, rather than supplanted, Taylor's focus on efficiency. This chapter does not provide an exhaustive literature review of this subsequent work. Sundstrom's recent (1986) book *Work Places: The Psychology of the Physical Environment in Offices and Factories* provides a good historical overview and summary of current research. This chapter's purpose is threefold: (1) to trace some of the principles or values underlying workplace plan-

Franklin Becker • Department of Design and Environmental Analysis, College of Human Ecology, Cornell University, Ithaca, New York 14853.

ning and design as these have evolved since the turn of the century; (2) to identify critical challenges facing practitioners responsible for making and managing workplaces and suggest areas where research could significantly contribute to improving the practice of planning, designing, and managing the workplace over the next decade; and (3) to illustrate with a case study how existing research can help guide the planning and design of the workplace.

HISTORICAL OVERVIEW

Research into the practice of planning, designing, and managing the workplace can be viewed as driven by a historically evolving set of broad attitudes and values (Becker, 1981, 1986, 1988d). These form the context within which researchers and practitioners frame the questions they ask and the space policies and physical designs they generate.

1910–1940: EFFICIENCY AND INDIVIDUAL PERFORMANCE

The dominant image here is the factory assembly line. Industrial growth was explosive. Stimulated by new communication and transportation networks, the family-run business that was located in a single area, if not building, gave way to huge industrial firms spanning large geographical areas and employing thousands. Management styles that worked for the craftsman, the self-employed professional, and the small administrative staff could not cope with management of large firms. Taylor's book heralded in the age of management.

The overriding concern of management was to improve organizational profits by maximizing individual task performance and efficiency (Barnes, 1949; Viteles & Smith, 1941). The worker was hired to do a specific task and was viewed as largely motivated by financial incentives (Taylor, 1911). Work was viewed as a sequence of specific tasks linked by formal organizational structure, not as a social process (Barnard, 1938). Little attention was paid to the effect of different layouts on communication processes, and how these in turn might affect individual performance. Even less attention was paid to workers' comfort and satisfaction (Barnes, 1944).

Human–environment relations-related research, which was essentially the study of human factors, was typically conducted in laboratories using highly artificial situations to test the effect of noise, for example, on different mental and physical performance tasks designed especially for the laboratory (cf. Hovey, 1928, cited by Sommer, 1967;

Viteles & Smith, 1941). Such research ignored the social relations that structured human activity and perception in the actual workplaces. The human relations movement evolved in response to this asocial view of work and the worker.

1940–1950: Task Performance and Social Relations

The first significant challenge to Taylor's ideas came in the form of the human relations movement and the famous Hawthorne studies (Roethlisberger & Dickson, 1939). Performance as an overriding concern was still paramount. The new twist was that social relations among workers, not merely financial incentives and formal organizational structure, were viewed as a critical factor affecting task performance (Berkowitz, 1954; Rice, 1958; Schachter, Ellertson, McBride, & Gregory, 1951). Documentation of how workers banded together in informal work groups to set their own production quotas, often by deceiving the time-methods analyst about how much work was really possible in a given amount of time, led to a view of work as part of a social process (Whyte, 1948).

This work, like that before it, was centered on factories, not offices, and was done very much from a management rather than worker perspective. In fact, the famous Western Electric studies by Roethlisberger and Dickson (1939), in which changes in lighting levels and the introduction of Musak had no *direct* effect on task performance, were widely interpreted as demonstrating that the physical environment of work had no effect on performance. This study effectively exiled the role of the physical setting from organizational studies. This is despite the fact that as Homan (1950) pointed out,

> men were working in a room of a certain shape, with fixtures such as benches oriented in a certain way. . . . These things formed the physical and technical environment in which the human relationships within the room developed, and they made these relationships more likely to develop in some ways than in others. (pp. 80–81)

Nonetheless, research on the role of the physical environment on communication patterns did not emerge until the 1950s and 1960s, and even then its influence on management studies was minimal.

1950–1960: Group Dynamics, Communication, and Conflict

One reason the study of what Sommer (1967) labeled as "small group ecology" had little influence on research in organizational behavior was that it paid scant attention to task performance in terms of traditional productivity measures. These studies were conducted by so-cial rather than industrial psychologists, and understanding group dy-namics, not maximizing organizational profit, was the driving force of this new human–environment research (cf. Hare & Bales, 1963; Som-mer, 1961; Steinzor, 1950; Strodbeck & Hook, 1961). It occurred outside the factory context and, in fact, outside of a management context.

Reducing interpersonal conflict and creating more democratic and cooperative group processes was the motivating force behind much of this work. Just as the rapid development of the railroads, telegraph, and then the telephone had created challenges which Taylor and his col-leagues met with scientific management, World War II and its horrific inhumanities stimulated social psychologists, in particular, to focus on group dynamics, conflict resolution, and attitude change. The parts of this work that explored the role of seating position (Sommer, 1961; Steinzor, 1950) and site planning and building design on communication and friendship formation (Festinger, Schachter, & Back, 1950; Gans, 1961) represented the beginnings of what evolved into a separate field of study within psychology: environmental psychology.

Not only were its sponsors different, but its audience also shifted. Stimulated by research such as Festinger *et al.*'s (1950) studies of how social norms and friendship patterns were influenced by the layout of student housing at MIT, architects and planners were beginning to be viewed as a more relevant audience than human resources managers. Environmental psychologists turned to the design professions as natu-ral allies in the quest to improve group and individual functioning as it was affected by the design of the built environment. For many design professionals stung by the highly publicized failures of designs to serve their users well, this was a welcome collaborative effort.

1960–1970: Focus on the Nonpaying Client

In contrast to human factors and human–environment relations studies done within an organizational framework, environmental psy-chologists turned to settings such as housing and schools. The emphasis here was on the end-user. Particularly in housing studies, performance was measured in terms of user comfort, satisfaction, and the ability of the environment to support desired lifestyles (Becker, 1974; Cooper, 1975; Francescato, 1979). Zeisel (1974) distinguished between the "pay-ing client," persons responsible for authorizing payment for physical design changes, and the "nonpaying client," the end-user who occupied and used the built environment. The former often had a great deal of influence over the form the built environment took; the latter typically had very little.

Working with nonpaying clients who typically had little voice in design decisions affecting their lives, the issue of user involvement also began to surface as a major theme for environmental psychologists and design professionals concerned with how the built environment affected human behavior (Alexander, Ishikawa, & Silverstein, 1975; Becker, 1974, 1977; Sommer, 1972, 1974, 1983). The fact that much of the research at the time focused on housing, widely viewed as the last bastion of choice and independence in a tightly controlled world, made the issue of the nonpaying client's involvement more poignant.

End-user participation in the design process ranged from providing information about work, living patterns, and environmental preferences to actually participating in the design process (Becker, 1977). Such participation was viewed as good (if not a right) *per se* in a democratic society (Alexander *et al.*, 1975). Research also suggested that user participation would lead to designs that more accurately reflected the needs, values, and preferences of the people occupying buildings (Wandersman, 1979). Thus it had an instrumental as well as political value, since it was likely to increase user satisfaction and reduce crime, vandalism, and maintenance costs (Newman, 1975).

The focus on housing and school settings (Coates, 1975), with its emphasis on user satisfaction and involvement, began to decline in the late 1970s, stimulated in large part by drastic reductions in government funding for research on public housing and schools. At roughly the same time, rapid growth in the service economy, with the associated increase in office workers and office buildings to accommodate them, began to focus attention on the workplace. The widespread introduction of computers into the workplace, viewed by many organizations as critical for dealing with increased global competition and the need to contain costs while increasing productivity, pushed the workplace even further into the spotlight.

It was the consequences of the introduction of the German *bureaulandschaft* planning concept into workplaces in the United States, first at Kodak in Rochester, New York, in the late 1960s (Brooks & Kaplan, 1972), that first attracted the attention of the environment–behavior field (Duffy, Cave, & Worthington, 1976). While the settings had shifted, the issues of communication, comfort, employee satisfaction, and user involvement were carried forward. For the first time, they were beginning to be integrated into thinking and research about the workplace from an *employee* as well as management perspective.

1970–1980: Communication, Worker Comfort, and Satisfaction

Bureaulandschaft, which came to be known in the United States as *office landscaping* (and then open planning), was based on a factory metaphor. Managing the flow of information as it entered the workplace, was processed, and left was viewed as a process comparable to the way in which materials flowed through a factory. Just as industrial psychologists had shown that factory layout could positively influence efficiency (and thus profits), the layout of offices was once again planned to expedite the flow of the knowledge worker's material—information—as it was processed on the way to becoming a product. The nature and flow of communication became a legitimate organizational outcome.

Rigid, enclosed offices with floor-to-ceiling walls were replaced with free-standing, easily moved panels about five-feet high. The goal was to arrange these in a manner that mirrored and facilitated the flow of information (typically in the form of paper) through the office. Using office size and degree of enclosure to communicate status was viewed as hindering the flow of communication and as undemocratic. The widespread adoption of open office planning in American offices provided a new impetus for environmental psychologists to study the workplace.

From the beginning, the open-plan office was resisted by American workers at all levels (cf. Becker, 1981). Visual and aural privacy were weakened, especially for professionals and middle level managers (Brooks & Kaplan, 1972; Justa & Golan, 1977). Contrary to the expectation that the open plan would enhance internal communications, many studies showed that communication declined or became almost entirely social (Clearwater, 1980; Oldham & Brass, 1979). Complaints about noise and loss of privacy were common (cf. Becker, 1981; Sundstrom 1986).

While the workplace was now the focus of much human–environment research, it was almost entirely different in character from the human factors-oriented research begun in the early part of the century and carried on with much the same values driving it ever since. First, research on the workplace was now firmly located in the office rather than the factory, reflecting a declining manufacturing sector and burgeoning service economy. Second, it focused much more on employee comfort, satisfaction, and communication than highly specific task performance. The nature of white-collar "knowledge" work made identification of clearly delineated task outcomes more difficult. In part, the relative disregard of traditional productivity measures could also be seen as reflecting the concern carried over by environmental psychologists from the research on housing where end-user comfort and satisfaction were considered worthwhile outcomes in their own right.

Management had never, however, reduced their concern for measuring task performance or reducing operating costs. High levels of employee dissatisfaction with open-plan offices made management aware of physical design issues, but they wanted harder evidence showing that such designs actually undermined performance, especially since open planning reduced the amount of space allocated per person, thereby directly significantly cutting operating costs (Becker, 1981).

Widespread publicity about employee resistance to office automation, including (often anecdotal reports of) high rates of absenteeism and turnover, came closer to the management mark (Makower, 1981; Working Women, 1981). Studies and reports linking resistance to computers to design factors such as uncomfortable lighting, ventilation, and seating persuaded management that outcomes such as comfort or satisfaction had significant organizational consequences, i.e., they were likely to cost the organization money (Briscoe, 1983, Caker, Hart, & Stewart, 1980; Dainoff & Happ, 1981). Research and speculation about how the introduction of computers would affect communication and supervisory patterns, by making information more widely available and allowing employees to work at home without traditional supervision, contributed to an interest in how office automation might affect supervisory patterns and management practices in the workplace (Becker, 1986; Leduc, 1979; Walton & Vittori, 1983).

THE PERFORMANCE PROFILE CONCEPT

In effect, themes and values which had ebbed and flowed in human–environment research over the past 75 years had begun to coalesce. Narrow concerns for productivity based on a factory metaphor were shifting toward a performance profile approach to assessing the effects of how offices are planned, designed, and managed (Becker, 1986, 1988a, 1988b).

The key to the performance profile concept is that any intervention is likely to have many different consequences. Some of these are likely to be positive, some negative. The challenge is to identify the overall pattern of outcomes rather than concentrating on one or two outcomes that may be relatively easy to measure (e.g., employee satisfaction) or have instant face validity (e.g., reduced operating costs).

Michael Brill's (1984) study of office workers represented in many ways the merging of different values; in particular, user involvement, comfort, satisfaction, communication, and task performance. Brill explicitly linked self-reported measures of satisfaction and comfort to what he called "bottomline" performance measures: job and environmental satisfaction and job performance. He argued that such organizational performance measures are meaningful because they have been empirically related, in other studies, to employee absenteeism and turnover. These, in turn, have direct and significant cost implications.

Thomas Allen's (1976) work in R&D organizations even more directly links communication, design, and performance. Allen relates the quality of engineering design solutions, as measured by panels of experts, to the amount and nature of communication among engineering groups during design development. He then looks at the effect of architectural design variables, particularly physical distance, on communication patterns. Productivity was not measured in these studies in terms of a traditional industrial input–output ratio, but performance has been linked to what Becker (1981) called "culturally powerful" measures. These are outcomes that within any given organization are highly valued. They are likely to differ from organization to organization, and even for the same organization over time (Steers, 1976). In effect, such outcomes acknowledge the organization's commitment to measurable performance indicators, but they assess these in a more sophisticated manner than the early input–output models which ignored what McGregor (1960) called "the human side of enterprise."

Summary of Historical Overview

The principle audience for much of the early human–environment relations work in the workplace, including time–motion and human factors studies, was the human resource manager and industrial psychologist. Individual task performance was the primary outcome measure. As interest by social and environmental psychologists and design professionals in how the planning and design of nonwork settings such as housing and schools increased, the audience shifted to embrace design professionals. Typical outcome measures were likely to include user comfort, satisfaction, and interaction patterns, as well as users' preference for and response to different forms of involvement in the planning and design process. As human–environment relations researchers began to focus on the workplace, albeit considering a broader range of outcome measures, the audience for human–environment research once again became management, but this time management with responsibility for the planning, designing, and managing of an organization's physical assets: facility management.

FACILITY MANAGEMENT

Like the performance measures themselves, facility management has evolved over the last 8–10 years as a function that integrates concern for performance and quality of work life within the organization. It is a function that, by definition, is concerned with people, place, and performance. At the beginning of the 1990s, it is the focal point for much of the energies surrounding research and thinking about how offices are planned, designed, and managed.

DEFINING THE FIELD

Becker (1987) has defined facility management as the field "responsible for coordinating all efforts related to planning, designing, and managing buildings and their systems, equipment and furniture to enhance the organization's ability to compete successfully in a rapidly changing world" (p. 82). There are three key ideas in this definition. First, facility management is responsible for the total design process, from planning through managing a building in use over time. Second, because facility management is responsible for both initial planning and managing the continuous evolution of buildings, furniture, and equipment over time, it must be concerned with the planning and design process, including the ways in which employees are involved in workplace decisions. Third, enhancing the organization's ability to compete successfully in a rapidly changing world means that the facility manager must think strategically as well as operationally. The facility manager must be aware of fundamental trends in operating environments, the labor force, new technologies, and economic and social conditions in order to best allocate the organization's physical resources to enhance performance.

KEY ORGANIZATIONAL TRENDS

A brief listing of some of the key trends (Becker, 1986, 1988c; Davis, Becker, Duffy, & Sims, 1985; Naisbett, 1982) facing organizations today begins to set in context the kinds of research and design challenges researchers and practitioners will face from now until the end of the century. These include:

- A more demanding workforce with higher expectations concerning office accommodations, health, and safety.
- More information technology requiring more sophisticated buildings and building services.

- More global competition requiring more dynamic and creative organizations.
- More complex problems requiring more creativity and varied expertise.
- A need for highly qualified professionals in short supply creating pressures in organizations to satisfy the individual's need for personal and professional identity, desire for challenging jobs, and greater personal autonomy.
- Rising costs of property and building in major cities stimulating cost containment measures while maintaining, or even increasing, existing service levels; that is, a concern for getting "value for money."

This list, while not exhaustive, suggests a set of conditions that presents the facility manager with enormous challenges. Seen as a problem of resource allocation, the overarching challenge from a facility management perspective is to contain facility costs while adding value. It is this objective that underlies several key issues related to planning, designing, and managing the office environment over the next decade.

FIVE KEY WORKPLACE ISSUES

While it is impossible to identify every issue of relevance to the planning, design, and management of the workplace in a single chapter, the following broad areas seem likely to present the major challenges to both researchers and practitioners concerned with improving the workplace over the coming decade:

- Control
- Communication
- Environmental change processes
- Performance
- International influences

Each of these issues is briefly discussed below, with the emphasis on critical unanswered questions.

CONTROL

Control is a recurring theme in both the management and human–environment relations literature, relating particularly to environmental stress from noise and crowding (Baum & Valins, 1979; Glass & Singer,

1972; Sherrod, 1974) and thermal comfort (Hedge, 1989); space planning and furniture layout (Becker, 1981; Froggatt, 1985; Worthington & Konya, 1988); and environmental change (Becker, 1981, 1986; Becker & Syme, 1987; Lynch, 1972). Arguably, it is the predominant issue facing researchers and practitioners alike, since it affects so many fundamental aspects of the working environment.

The workplace literature is consistent in showing that American workers want more involvement in decisions affecting the planning and design of their workplace than they actually have (Brill, 1984; Harris and Associates, 1981, 1987). Several studies have also shown that higher levels of employee involvement in environmental planning and design processes are associated with higher levels of environmental satisfaction (Becker & Syme, 1987; Brill, 1984; Froggatt, 1985). A study of employee involvement in decision making (DIO, 1983) found that increased employee involvement was not necessarily more time consuming than more autocratic decision processes. This study also found greater employee commitment to decisions made with higher levels of employee involvement. Sommer (1983) reports a study in which employee involvement in the design process resulted in fewer change orders, thus saving the organization both time and money in a major building project.

What is not clear from research on user involvement to date are the following:

1. The effects of different *types of involvement* on performance. Froggatt (1985) found that employees who selected their own office furniture package were more satisfied with their offices than those who at a later point were not given this opportunity. The effects on other performance measures were not examined. In a dormitory setting, Wandersman (1979) found that students preferred working with a designer to design their dormitory room, but they wanted to retain ultimate control over the final design decision. Several typologies of user involvement have been developed (Froggatt, 1985; Wandersman, 1979), but little research has systematically related different types of user involvement to a range of outcome measures.

2. Research is lacking on the *stages in the planning and design process* at which different kinds of employee involvement have the greatest positive benefit in terms of performance measures such as environmental satisfaction, job satisfaction, job performance, and motivation. There is little research such as the DIO study (1983), which found, for example, that employee involvement in the problem definition stage of decision making was associated with more employee commitment to and satis-

faction with subsequent decisions. Employee involvement in implementation stages, which is more typical, was associated with lower levels of
commitment and greater inefficiency.

3. What types of *environmental control* is most preferred by employees, and whether this varies as a function of individual difference
variables such as job level and job function, gender, or previous work
experience. Several studies, not directly concerned with control issues,
suggest differences in response to environmental conditions by gender
(Hedge, 1989), job level (Becker & Syme, 1987; Froggatt, 1985; Sundstrom, Burt, & Kamp, 1980), and previous work experience (Becker &
Hoogesteger, 1986; Becker & Syme, 1987). Some have shown greater
similarities than expected, especially in terms of privacy requirements
(Sundstrom *et al.*, 1980). In terms of environmental control, much of the
physical design of the workplace has focused on and provided
ergonomic controls for seating, lighting, and to a lesser extent tem-
perature, ventilation, and furniture selection and layout. The relative
importance to employees of such forms of environmental control compared to, for example, designing whole workstations, being able to rear-
range one's furniture within the office, select and locate different
amenities (e.g., break areas, fitness facilities), or select the style of furniture is not well understood.

4. Research has not explored whether employee preferences for the
*types of involvement, type of environment control, and stages at which these
occur* (1) differ as a function of employee variables such as age, gender,
job level, job function, education, or stage in one's career and (2)
whether these individual difference factors affect performance
measures.

5. The role of planning and design processes as a form of *organizational development* is poorly understood. Can certain forms of user involvement be used as a device for clarifying fundamental organizational
objectives, for resolving intergroup conflict, improving communication,
or reducing stereotypical thinking among different disciplines and departments? Steele (1973, 1977, 1986), Becker (1981, 1986), and others
(Worthington & Konya, 1988) suggest that planning and design processes can have a broad impact on organizational development efforts,
but little systematic research exists.

6. Little systematic evidence exists showing whether certain forms
of user involvement affect *business performance* directly, e.g., by reducing
operating and maintenance costs or reducing change orders. In a review
of Richard Rogers and Partners Lloyd's of London building, Becker
(1988b) found that failure of the design process to effectively incorporate
end-users views resulted not only in considerable user dissatisfaction

but in considerable sums of money to modify one of the most expensive buildings in Europe 18 months after initial occupancy.

COMMUNICATION

The literature on the physical environment and communication has shown that distance affects the frequency of (undefined) face-to-face communication (Allen, 1976; Festinger *et al.*, 1950). Allen's work shows, for example, that in engineering organizations the frequency of communication falls off sharply beyond about 150 feet. Sommer's (1969) early studies of small group ecology show that different types of seating arrangements are preferred for different kinds of interaction: side-by-side for cooperation, kitty corner for coaction, and across the table for competition.

A number of studies have also shown that physical environment conditions affect interpersonal attraction and impression formation. For example, Maslow and Mintz (1956) found that ratings of male and female photographs were higher in a beautiful room than in an ugly room. Baum and Valins (1979) found that college students living in dormitories in which it was difficult to control interaction with other students were more likely to withdraw from social situations outside the dormitory than students living in dorms whose designs made it easier to control interaction. Sauser, Arauz, and Chambers (1978) found in a simulation study comparing interpersonal judgments under quiet and noisy conditions that subjects assigned lower starting salaries to persons in a noisy room than in a quiet room. A number of studies have examined the effect of seating position in offices on impression formation. In general, seating arrangements that place furniture as a barrier between conversants are perceived as more formal and less open than arrangements in which two people sit with no physical barriers between them (Campbell, 1979; Morrow & McElroy, 1981; Zweigenhaft, 1976).

All of these studies are relevant for those who plan and design office settings. Much of the research has focused, however, on nonverbal communication. With respect to direct, face-to-face communication a number of issues are worth examining.

1. *The physical environment and different types of communication.* As with "user involvement" a major problem in the literature is the failure to distinguish among different types of communication. Allen, Becker, and Steele (1987) distinguish among three types of face-to-face communication (see Figures 1–3). These are:

a. *Coordination*, which refers to the sharing of information necessary to coordinate the efforts of different departments or work groups. This communication is usually scheduled and formal, and it usually takes place in designated meeting places such as conference rooms. It is the most typical form of organizational communication and the kind most deliberately designed for.

b. *Informative*, refers to the sharing of information necessary to remain current in one's professional field. It is critically important in rapidly changing technical fields such as engineering, science, and medicine, but it is important for any employee working in an area where ideas, technologies, and procedures are changing. Resource centers and libraries are the visible organizational recognition of the need for this form of communication. Allen's (1976) research shows that, at least among engineers, information of this sort is more often shared by word of mouth. Thus it is likely to occur in virtually any location, including offices, corridors, cafeterias and break areas, fitness centers, and the like.

c. *Inspiritional*, refers to the generation of new ideas, creative thinking, and imagination. Of the three forms of communication, it is the most random and unpredictable type. It is often unplanned and serendipitous. Allen's research, again, suggests that it is enhanced by fre-

FIGURE 1. Communication to coordinate plans and programs is critical in all organizations. In large organizations this is done typically through scheduled meetings in conference rooms.

FIGURE 2. Communication to inform is a means of staying abreast of developments in one's field. It is essential in professional and technical fields that experience rapid change in their knowledge base.

quent face-to-face contact with persons *outside* one's own work group. It is the form of communication that buildings and interior layouts are least often deliberately designed to accommodate.

Historically, to the extent it is deliberately considered at all, design related to fostering inspirational communication has been associated with R&D units (Allen, 1976) because their manifest objective is to innovate. Arguably, such concerns are now as much a part of organizational life in once staid industries such as banking and insurance. Worldwide

FIGURE 3. Communication to inspire is required wherever new ideas are being developed. This form of communication, which tends to be more serendipitious, is the least well accommodated in most office planning and design.

competition and deregulation of these industries have generated enormous pressures for innovative "products" in the form of consumer services. Thus, these as well as many other types of organizations that must innovate to remain competitive, such as utilities and telecommunications, are likely to find all three forms of communication critical to their survival.

2. The role of microdesign features. In addition to not distinguishing among different types of communication, little work has examined environmental factors other than proximity that affect communication. Steele (1986), based on his observations as an organizational consultant, suggests that microdesign features such as the location of circulation routes, water fountains, mail distribution centers, copy machines, or rest rooms act as activity generators or "behavioral magnets" that make it more likely that people will go in one direction than another. The early work on housing layout and social patterns (Festinger *et al.*, 1950) supports these ideas.

There is little specific research on the workplace that has examined the effect of such microdesign features on communication patterns. Nor has the effect of physical characteristics of designated communication settings such as cafeterias, break areas, or conference rooms received much systematic examination. Some of the research on interpersonal attraction and impression formation, noted above, suggests that the fit, finish, lighting, and acoustic characteristics of meeting areas can affect nonverbal communication.

3. Ratios of type of meeting areas. Even more fundamentally, there are no systematic data to guide designers in establishing the ratio of different kinds of communication spaces (e.g., conference rooms) to number of employees as a function of factors such as type of work, building layout, or the design of individual workstations. In a study of the spatial ecology of commons areas in a small R&D firm, Techau (1987) found that use of commons areas was significantly affected by distance factors. Areas designated for general use were most likely to be used by groups most proximate to them. This suggests that the number of conference rooms needed may be as much a function of their location (and therefore their use patterns) as of their capacity in relation to overall staff size.

4. Role of dedicated project rooms. Little research exists on the effect of different designs and locations of special-purpose places, such as dedicated project rooms, on the frequency or nature of communication, cohesiveness, and the like, for project groups, or their relation to other colleagues and supervisors outside the project team.

5. *Centralization versus decentralization.* Still another unexplored area in an organizational context is the effect of spatial centralization and decentralization policies. Allen's (1976) work in R&D firms suggests that decentralization is likely to significantly reduce communication, even beyond several hundred feet, two or more vertical floors, or two adjacent buildings. Of particular interest is how nonphysical factors, such as policies governing the frequency of movement between remote settings, or new technologies such as electronic mail and bulletin boards, mediate physical distance and affect communication and supervision patterns, group and corporate identity, and employee commitment.

6. *Open vs. closed offices.* The question of the relation between open offices (i.e., an individual work area not enclosed by floor-to-ceiling walls) and closed offices (i.e, an individual work area enclosed by floor-to-ceiling walls) should be examined in more detail. The issue is not whether one or the other is preferred (an early question) but what is the appropriate relationship between not only these two types of offices, but among these two types and other spaces within the organization.

7. *Multiple work settings.* Becker (1986) has emphasized the importance of thinking of the office as a series of loosely coupled settings. These may be linked electronically, by conventional mail, or by personal movement among workers. An interesting question is how to design and manage the movement of people and information among these settings. As the case study below suggests, provision of multiple settings involves issues of organizational culture and management policy as much as physical design. The relationship among these overlapping domains has received little attention.

ENVIRONMENTAL CHANGE PROCESSES

Despite the fact that change is endemic in American society, little research has examined environmental change processes in office environments (Becker & Hoogesteger, 1986; Becker & Syme, 1987; Zube & Sell, 1986). Research has focused on issues of residential mobility (Michelson, 1977) and environmental meaning, especially in relation to historic preservation and urban design (Lynch, 1972).

Conventional wisdom argues that people almost always resist change. Public outcry against urban renewal projects (Fried & Gleicher, 1961; Gans, 1968), new highways, and commercial development (Gibson, 1981, cited by Zube & Sell, 1986) provide ready examples. Yet such resistance to change seems paradoxical in the context of the enormous interest in leisure, including travel. In the context of *voluntary* choice, people seem eager for change.

The key question, of course, is the *conditions under which* people do or do not resist particular kinds of environmental change. Equally important is how people use aspects of their physical surroundings to help them cope with changes in their social circumstances, such as marriage, aging, transition into a new job, school, or community (Cohen *et al.*, 1986). Weick (1978) argues that people need both stability and change. What becomes critical is the pattern of these complementary relations. These have a time as well as a spatial dimension (Lynch, 1972).

Becker (1981) related Lynch's temporal dimensions to the work environment. They included *grain* (the size and precision of the chunks into which time is divided), *period* (the length of time within which events recur), *amplitude* (the degree of change within a cycle), *rate* (the speed with which changes occur), *synchronization* (the degree to which cycles and changes are in phase), and *regularity* (the degree to which the preceding time characteristics themselves remain stable and unchanging.

In a recent article, Zube and Sell (1986) note that little research on environmental change processes has been longitudinal. Becker's (1986, 1987) recent work has begun to address these issues. In a pre–post study of government employees' response and adjustment to an office relocation from downtown Washington, D.C., to suburban Maryland, Becker and Hoogesteger (1986) found that few employees rated adjustment to this type of office relocation very difficult. Somewhat surprisingly, middle-aged employees had significantly more difficulty adjusting than those older or younger. Employees who had never moved before, or who had moved many times before, adjusted more easily than those between the two extremes. Managers, who typically are the most informed about a move, were more satisfied and adjusted more easily than clerical and technical/professional employees. Ironically, after moving from a rather dreary office to a new, freshly painted and carpeted one with modern systems furniture, better lighting, and the like, employees rated their overall satisfaction with the physical aspects of the new workplace at the same level as their previous one.

A second study focused on the kinds of information provided by management that employees felt would ease their adjustment to an office relocation and the aspects of the physical environment that were most difficult to adjust to (Becker & Syme, 1987). In this study the relocation involved moving from single tenancy in a renovated high school on the edge of a small New England town to shared tenancy in a mammoth modern office building in the middle of over 600 acres of woodland 10 miles away.

Employees reported wanting more information about the environment at the workstation than at the building level. This was true both before and looking back, after the move. As in the earlier study (Becker & Hoogesteger, 1986), self-reported adjustment to the new building was generally easy. Ease of adjustment at the building scale did not change over time, from just before the move to several months afterwards. Actual adjustment to workstation issues was significantly easier than expected. Unlike the first study, upper management anticipated the most difficult time adjusting and, while actual adjustment turned out to be easier than expected for all groups, it was more difficult for management. Satisfaction with the new office was strongly correlated with actual ease of adjustment. Both studies contradict the conventional wisdom that people resist and have a difficult time adjusting to change. These two small studies suggest that people have a difficult time adjusting to change that undermines their personal and professional identity and sense of competence.

While these studies mark a beginning of deliberate examination of environmental change processes in the workplace, they beg more questions than they answer. These include:

1. What is the *length of the adjustment period* to different kinds of physical change, and how do individual difference factors affect this adjustment period? Is a stage theory of adjustment appropriate, and if so, what are the relevant stages and their characteristic behavioral patterns?
2. What is *the nature of adaptation,* including the role of the physical environment and social networks, in making adjustment to change easier or more difficult?
3. What role do *formal organizational support processes* and events play in the environmental adjustment process? What is the relation between these formal intervention efforts and informal processes among colleagues, friends, and family?
4. What are the most *effective media* with which management can communicate with staff: video, site visits and open houses, photographs, newsletters?

Performance

For applied researchers and practitioners interested in improving the quality of the work environment, understanding performance issues is mandatory. As noted earlier, without considering the relation between environmental change and performance, the likelihood of being able to influence physical decisions in organizations in severely limited.

Supported by a National Science Foundation grant, the American Institute of Architects recently held a conference on "The Impact of the Work Environment on Productivity" (Dolden & Ward, 1986). The discussion of productivity, for the most part, reflected the kinds of measures associated, above, with the period roughly corresponding to the early 1980s; that is, self-reported measures of employees' satisfaction with aspects of their physical work environment, and to a lesser extent, with their jobs. Survey research was the most common methodology used. In fact, few of the studies presented or available in the literature directly assess the effects of environmental change efforts on performance of any sort. Several researchers have attempted to extend self-reported performance measures to include more culturally powerful indicators, namely, money and product quality (which is assumed to translate fairly directly into financial benefits).

Brill's (1984) work assumes that the bottom line for most organizations is financial. He therefore applied a controversial method of converting self-reported measures of job satisfaction and performance into (hypothetical) dollar savings. Sommer (1986) also has proposed translating social costs into financial figures as a way of attracting the attention of organizational decision-makers and dealing with them in terms with which they are familiar. Allen (1976) has used panels of experts to rate the quality of design solutions. He then related these ratings to the frequency of communication within and between work groups.

Becker (1986) has distinguished building performance (e.g., ease of maintenance, capacity to accommodate new information technologies, and ability to respond to organizational and technical change) and human performance (e.g., environmental satisfaction, quality and quantity of output, job satisfaction, and performance ratings). Whether for building or human performance, any environmental intervention is likely to have multiple outcomes.

The concept of the "performance profile" (Becker, 1986, 1988d, 1988e) is a means of acknowledging that any intervention stimulates a wide range of responses, some positive some negative, some observable, others subjective but recordable. Judgments are made taking into consideration relative weightings of different outcomes. These may differ across organizations, or for the same organization over time (Steers, 1976).

Critical unresolved issues related to the planning, design, and management of the workplace and performance are:

1. The nature of outcome measures valued by different kinds of organizations, under different kinds of market and labor force conditions. For what kinds of organizations, or decision-making styles, or types of organizational culture are quantitative or qualitative data most valued? What are the best ways of presenting these data to decision-makers?

2. The need to reconsider the nature and range of the methods used to assess performance. How can existing archival data, often collected electronically, be exploited to measure performance? Spatial mapping of absenteeism, turnover, employee satisfaction, and the like might easily be accomplished using human resources data integrated into facilities management databases. Such data, mapped over time and multiple locations would begin to generate patterns of human–environmental relations unable to be identified from small, one-off studies of single departments or small groups. While direct observations are time consuming, the kinds of field experiments done with small group ecology by Sommer (1969) years ago still provide a useful model.

INTERNATIONAL INFLUENCES

Multinationals operate worldwide. The huge commitment to establishing office and manufacturing facilities in foreign countries has stimulated interest in understanding how the kinds of issues identified above are affected by different cultural influences. Research in this area is in its infancy, but it is beginning to accumulate.

Ekuan (1982), for example, compared employee response to offices in Los Angeles and Tokyo. He found, not surprisingly, less privacy and personalization and higher density in Japanese offices than in American offices. Space and furniture allocation by rank, typical in most American offices, was minimal in Japanese offices.

In a study focusing on office quality and office procurement and development practices in Britain, Germany, and the United States, several differences across national boundaries emerged (Ellis, Becker, & Jockusch, 1987). In Germany, as in other northern European countries such as Sweden and the Netherlands, there is a statutory right of workers to participate in decisions affecting their workplaces, and a much lower proportion of speculatively built offices. This contrasts strongly with the United States, where workers have no statutory right to participate in workplace decisions and management typically exercises firm control over the planning and design of the workplace. Britain tends to fall somewhere in between, without statutory provision for user involvement but a weaker corporate hand in planning and design decisions.

In both Germany and the United States, the quality of the office environment, both for employees and as an expression of corporate image and identity, is much stronger than in Britain. Unlike Britain and Germany, in the United States much more flexible leasing provisions, including leasing the bare shell with the tenant doing his own fit-out and leases of all durations, combine to create facilities that are flexible and meet organizational requirements, even in an office market dominated by speculative office buildings.

In a study of critical issues in the planning and management of foreign facilities facing six leading American multinational electronics firms, Becker and Spitznagel (1986) found that key problems centered around organizational relationships, including organizational structure and the consequent frequency and nature of communication across departments and disciplines, as well as regions, countries, and continents. Specific issues included the link between corporate and facility planning, the nature of approval processes for site acquisition and leases, new construction and major renovations, what to standardize, how to communicate and enforce whatever standards are adopted, training of professional staff (including development of technical skills and understanding of corporate culture and philosophy as they affected facilities decisions), and the nature of appropriate property strategies (i.e., leasing vs. ownership) under different market and national conditions. The report concluded that organizations need to distinguish more clearly between *strategic* and *operational* decisions. Typically, more effort is now devoted to controlling operational decisions (building standards), even though these have less financial and organizational impact than more strategic facility decisions (e.g., lease vs. own site location). Much effort was spent in justifying decisions rather than evaluating whether the expected costs and benefits matched the actual ones.

In a more focused professional report (DEGW, 1988) of different facilities requirements among multinational financial service firms with trading floors in their offices in New York, London, and Tokyo, overall space requirements, in terms of such factors as floor area, floor-to-ceiling heights, and density of occupation, were remarkably similar despite enormous cultural variation. This contrasts with less technology-driven office space. Becker (1988a) described a situation in which a major electronics firm's plans for a sauna in their Finnish office were rejected out of hand by corporate planners in the United States who were shocked at the attempt to include such "luxury" facilities in the offices of a firm dedicated to doing things in a simple, economic manner. Only with great effort and considerable time and energy were the Finnish nationals able to persuade their American counterparts that in Finnish society a sauna was a necessity, not a luxury. It is where contracts are negotiated and concluded.

Another professional study by DEGW (1987), based on focus groups with small numbers of accountants and solicitors in the city of London, found that in Britain office designs that make a strong "statement" are considered bad form, likely to alienate customers who are concerned that they are paying for unnecessary environmental frills rather than simply good service. Building services such as air conditioning, considered essential and largely taken for granted in Japan or the United States, were still viewed with suspicion by many British firms.

To date, cross-cultural studies concerning the planning, design, and management of the workplace have no common theme. Most are occurring in the context of facility planning and management. Not surprisingly, they suggest that national characteristics affect office planning and design. Important questions to be addressed in an international context include:

1. The advantages and disadvantages of different kinds of standardization programs, from exterior facade and signage to space and furniture standards.
2. The meaning of different levels of fit-out, particularly furniture and finishes, to staff and customers.
3. The range of typical work patterns, by industry type, and the kinds of physical accommodation considered appropriate to support them.
4. Attitudes to change, in general, and how these affect planning and design processes.
5. The role of national codes and building regulations on building form; and, in turn, how such building forms affect the organization's ability to accommodate new information technologies and to attract and keep a highly qualified work force composed of both nationals and foreigners.
6. The relation between worker expectations about control, supervision, and autonomy and space planning and building location and design.

THEORY INTO PRACTICE

The rather formidable list of research questions suggested above could occupy several lifetimes of research. In the meantime the world moves on and buildings are built, occupied, transformed, and used. Decisions are made, some terrible, but a lot quite sensible, with whatever information is at hand. Environment–behavior research done over the last decade has stimulated new approaches to planning and designing office buildings. Unfortunately, for the most part the practical applications of research data are embedded in unpublished projects. The Steelcase Corporation's new Corporate Development Center (CDC), with which the author has been involved in a consulting capacity, serves to illustrate some of the ways in which theory can be integrated into practice.

The effect of Steelcase's new Corporate Development Center on individual and organizational performance will not be known for some time. The principles guiding the planning process and physical design can be described, however. While the extent to which research and theory has been used in the CDC project is rather unique, for those who sometimes despair over integrating research into practice it provides a useful beacon of hope. For practitioners it provides a tangible, kick-the-tires model of planning and design processes and of a building and space planning project that they can examine, point to, and learn from.

The CDC translates theory into practice for planning processes as well as for physical design (Becker, in press). Steelcase, Incorporated, is the world's largest manufacturer of contract office furniture. To remain competitive, in addition to providing outstanding service for the furniture it sells, it must periodically introduce new products. A major motivating goal for the CDC was to explore how a new building, the immediate impetus of which was to accommodate growth that had exceeded the existing facility, could enhance product innovation and reduce the time involved in developing new products.

Improving communication among different disciplines and departments was a central consideration. Figure 4 differentiates between a "relay race" model of communication in the design process, in which information generated by one group or discipline is handed off to another group in a sequential process, and a "rugby" model of the product development process. In the rugby model, different disciplines and departments interact in a dynamic constantly fluctuating fashion. Leadership shifts as the nature of the project evolves. The hallmark is that all "players" are involved from the beginning.

The planning and design concepts used draw heavily on and integrate the work of Thomas Allen, Fritz Steele, and the author. Allen's work in R&D firms has shown, for example, that the amount of communication employees have with persons outside their own project team or department is correlated with outcomes such as better product design.

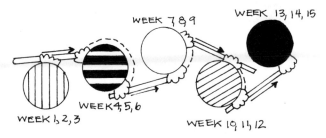

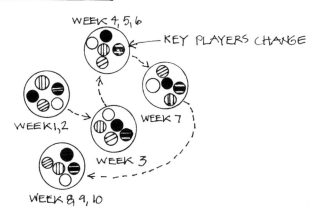

FIGURE 4. The relay versus the rugby model of planning for innovation.

As distance increases vertically and horizontally, such serendipitous contact declines. Steele (1986) has written about the symbolic aspects of design, and in particular, the need for top management not to isolate themselves from the rest of their staff, for both symbolic purposes and because of the effects such isolation has on the sharing of information and ideas up and down the line. Becker (1981, 1982, 1986, 1987) has written about the need to think of the workplace as a series of loosely linked settings in which any single workplace setting is unlikely to effectively support the full range of tasks and activities employees engage in over the course of a day, week, month, or life of a project. Both Becker (1981, 1986, 1987) and Steele (1986) have also written about the importance of involving employees in decisions affecting the planning and design of their workplace.

The concepts driving the work of Allen, Steele, and Becker have been expressed in the Steelcase CDC building process and design in several ways. These are briefly described below.

PROCESS: EMPLOYEE INVOLVEMENT

In addition to the typical involvement of senior management in all aspects of the project, the consultant team worked closely with a dynamic, talented, and enthusiastic project team from Steelcase to involve over 450 employees, all those who would occupy the new building, in the design process. Figure 5 shows the organizing framework the team used. A matrix is formed by linking decisions at different environmental scales, from the site and building to the workstation, to different levels of staff, from top management to rank and file workers. Decisions that affect the organization as a whole, such as the basic building form, are

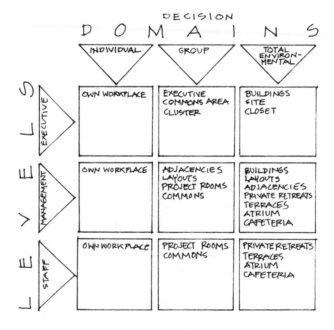

FIGURE 5. Opportunities in the design process. An environmental decision matrix creates a framework that suggests the kinds of decisions most appropriate for different groups within the organization.

made by top management. Decisions that affect the individual employee or work team are made by those most directly affected.

In practice, this has meant that site location, building form, and general design concepts were made by top management, based on concepts for stimulating communication and interaction and achieving internal planning flexibility suggested by the consultants research and experience. Decisions about the design of individual workstations, shared commons areas, and dedicated project rooms (see below) are made by involving virtually every employee in a series of intensive focus groups. These give employees the opportunity not only to describe their work patterns and environmental requirements, but also to react to proposed designs in early stages of the development process.

To generate realistic expectations, every attempt has been made to clearly articulate to all concerned the nature of decisions each group can influence, and why. For rank and file employees this included the nature of some building materials, such as glass.

For example, a full-scale section of the building was mocked-up on the new building site. An innovative new glass intended to reduce glare on VDU screens while permitting employees to see outside was installed in this test facility. Employee response convinced the designers to go back to the drawing board. Eventually, a new design was generated that worked better, saving thousands of dollars on glass that would not have met its intended objectives.

The planning process has used postoccupancy evaluation data (collected on the existing building), evaluation data from a test building, data from informal "wandering" interviews with staff, and feedback about preliminary designs in a series of focus groups to directly involve staff in decisions about their workplace. These are techniques that available research and literature, noted earlier, has consistently indicated will result in higher levels of employee satisfaction and designs that better meet employees and the organization's environmental requirements. These kinds of planning processes also increase different groups' interactions with and understanding of each other.

PRODUCT: DESIGN CONCEPTS AND SOLUTIONS

The organization concepts underlying the CDC project were given physical form through the following kinds of design characteristics.

Multiple work areas. Becker (1986b) used the concept of "loosely coupled settings" to describe multiple work areas, each designed to support particular kinds of activities and among which employees moved as their tasks varied over the course of time, whether a day,

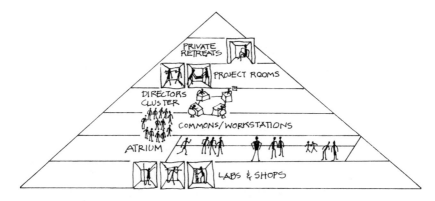

FIGURE 6. Multiple work areas reflect the fact that the same person engages in different tasks over time, some of which are better performed in space specifically planned for them.

week, or longer period. The concept of employee mobility is central to the CDC building (see Figures 6 and 7). Employees are expected to move among the following settings: individual workstation, shared common area near their workstations, dedicated project rooms, conference rooms, break areas and cafeteria, outside terraces, laboratories, and an atria serving as a kind of town square. There is also a sophisticated resource center ringed by small, fully enclosed offices that are used on a temporary basis for work requiring high levels of concentration. Many

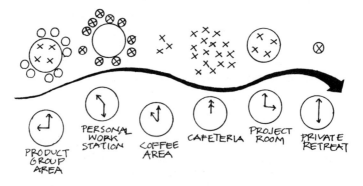

FIGURE 7. A variety of settings for a variety of tasks. Multiple work settings require a mobile worker who moves from setting to setting within the total workplace over the course of time.

of these settings will be linked electronically so that work begun in one can be continued after moving to another. The intent of the multiple settings is to recognize, as the research literature suggests, that neither open-plan offices nor fully enclosed offices *by themselves* support the full range of activities in which many employees engage.

Director's cluster. Rather than locating all senior management together at the top of the pyramid, in an isolated executive enclave, or to locate directors with their staff in departmental areas separated from each other, a director's cluster has been placed in the middle of the building. It is easily accessible via both stairs and escalators. Its central location increases the likelihood that staff on the way to the Resource Center at the top of the building, or to see other colleagues, will run into or at least have some visual connection to their senior management. The intent is to stimulate contact among the leaders of the various divisions involved in the product development process without isolating them from their staff.

Break areas and circulation. Break areas distributed throughout the building, as well as placement of mail, copy, and other frequently used support services along major circulation routes, are intended to stimulate serendipitous face-to-face contact. Escalators, which unlike elevators or hidden stairs allow people to see each other as they move about the building, are also intended to stimulate face-to-face interaction.

Adjacencies. Relative location of groups to each other, as well as different kinds of settings (e.g., project rooms and workstations), are linked to management and design decisions. Most adjacency decisions use spatial proximity to reinforce organizational structure. People are located nearest those with whom their job requires they frequently interact (see Figure 8). Allen's research suggests such arrangements are efficient, but not necessarily effective. The reason is that performance in R&D settings is related to the number of contacts one has outside one's own department, discipline, and project team.

If the goal is to enhance creativity, then stimulating face-to-face communication among persons whose jobs do not require interaction (weak organizational connection) is appropriate (see Figure 9). In general, those whose jobs require them to interact (strong organizational connection) will do so within reason. Thus, at the CDC, project rooms were located on different floors, away from individual workstations and departments. Efficiency (in its narrow sense) is reduced, but the expectation is that informal contact and communication will be increased, and it is this that has been shown to contribute to more innovative engineering designs.

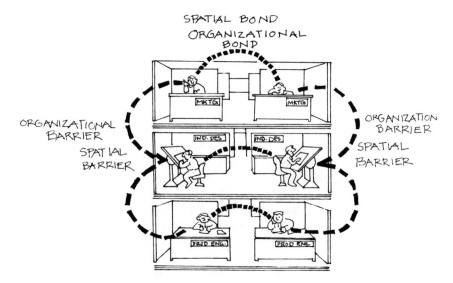

FIGURE 8. Adjacency requirements are typically driven by the desire to reinforce organizational bonds with spatial bonds. This limits informal communication among people who have a weak organizational bond but might benefit by sharing ideas and information.

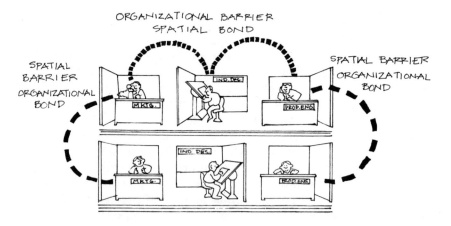

FIGURE 9. One way of overcoming an organizational barrier is to create a strong spatial bond. In this case spatial adjacencies are driven by a concern for which employees might benefit from talking with each other irrespective of their departmental or discipline affiliation.

The Acceptability Factor and the Enculturation Process

For anyone familiar with the typical planning and design of American workplaces, the nature of the risk–and potential benefit–at the CDC will be clear. Design innovations such as multiple work settings, work retreats, and (seemingly) inefficient adjacencies challenge traditional management styles and work patterns. To use such design opportunities to their full potential, employees from the president to the secretary need to fully understand the concepts behind the building. More critical, they need to know that if their supervisor sees them having coffee in the lounge at 2 p.m. that this will be perceived as working, not loitering; that if they remove their tie, they will not be viewed as sloppy or indifferent; that if they are not instantly available because they are in a private study area, that this will not be interpreted as being aloof or unresponsive (see Figure 10). Design innovation demands organizational innovation, and Steelcase realizes this. A year before occupancy an acculturation process is being developed that not only explains the concepts behind the building, but helps people develop attitudes and

Figure 10. Thinking counts as "real work." For informal break areas to be used, rank-and-file staff must believe that management will interpret their being in these areas as "real work," not a sign of their wasting their time.

behavior patterns that support the design concepts. Like the planning process, the acculturation process will involve all employees in a variety of face-to-face discussions and exercises.

Neither the perfect building nor the perfect planning process exists, and Steelcase's CDC is no exception. It is a useful model, however, to suggest how lines of research ranging from small group ecology and user involvement in the design process to facility management and innovation in R&D organizations can be applied toward making very specific decisions for buildings costing millions of dollars and affecting hundreds of employees.

IMPLICATIONS FOR FUTURE RESEARCH AND APPLICATIONS

There is an enormous reservoir of unanswered questions about how the planning, design, and management of the workplace affects individuals, groups, the organization, and ultimately the community and nation. Research could usefully address questions such as:

1. The nature and type of employee involvement that best fits the often competing demands of the individual and the organization. What kind of involvement is best for different types of organizations, different types of organizational culture, different types of employees? We need hard data on the time and cost associated with different forms of employee involvement. We also need information about the range of management styles and roles that best support forms of employee involvement that are valued by employees and beneficial to the organization.
2. The role and contribution of the "total workplace" (from workstations and atria to dedicated project rooms, resource centers, mail rooms, break areas, cafeterias, in-house stores, and shared common areas) in attracting and retaining employees, in stimulating informal communication, and in improving overall work effectiveness. How should such spaces be designed? Where should they be located? How many are needed? Who should control their use?
3. Alternative ways of designing and managing space that can accommodate unpredictable and high rates of organizational change. Is a highly changeable, flexible physical environment the answer? If so, which aspects of the overall design should be flexible, and which might be more effective if relatively static, or even fixed?

4. How can the physical environment help ease employee adjustment to organizational and technological change? Are there aspects of the individual work area that should move with the employee whose office is relocated? What policies and management practices can be devised that help make physical change in the office more palatable to employees?

5. What can we learn from other nations and cultures about office planning, design, and management that may help us address our ongoing concerns with accommodating organizational change, new technologies, shifting employee expectations, status, privacy, and informal communication?

The list of research questions will never be exhausted, or ever completely answered. Yet existing research, incomplete, contradictory, and limited as it is, can help improve the working environment, and perhaps help create more meaningful and rewarding working conditions. The linkages will, as always, be tenuous and indirect, the generalizations stretched too far, the number of decisions far outstripping the amount of information. Yet in wider time frames than a year, or five, or twenty, things do change. From 1911 to 1988, research and thinking on the workplace has moved from a total preoccupation with narrowly defined task performance to one that *adds* to it a concern for employee satisfaction, communication, comfort, and well-being. It had done so for reasons far flung from environment–behavior research, but also in small ways connected to and drawing from it.

REFERENCES

Alexander, C., Ishikawa, S., & Silverstein, M. (1975). *The Oregon experiment.* New York: Oxford University Press.

Allen, T. J. (1976). *The flow of technology.* Cambridge, MA: MIT Press.

Allen, T. J., Becker, F., & Steele, F. (1987). *The Steelcase CDC: Building for innovation.* Grand Rapids, MI: Steelcase, Inc.

Barnard, C. I. (1938). *The functions of the executive.* Cambridge, MA: Harvard University Press.

Barnes, R. M. (1944). *Work methods manual.* New York: Wiley.

Barnes, R. M. (1949). *Motion and time study* (3rd ed.). New York: Wiley.

Baum, A., & Valins, S. (1979). *Architecture and social behavior: Psychological studies of social behavior.* New York: Erlbaum.

Becker, F. (1974). *Design for living: The residents' view of multi-family housing.* Ithaca, NY: Cornell University Program on Urban and Regional Studies.

Becker, F. (1977). *Housing messages.* Stroudsburg, PA: Dowden, Hutchinson & Ross.

Becker, F. (1981). *Workspace: Creating environments in organizations.* New York: Praeger.

Becker, F. (1982). *The successful office.* Reading, MA: Addison-Wesley.

Becker, F. (1986a). Work in its physical context: The politics of space and time. In M. Dolden & R. Ward (Eds.), *The impact of the work environment on productivity*. Washington, DC: Architectural Research Centers Consortium.

Becker, F. (1986b). Loosely coupled settings: A strategy for computer-aided work decentralizations. In B. Star & L. L. Cummings (Eds.), *Research in organizational behavior*. Greenwich, CT: JAI Press.

Becker, F. (1987, November). New facilities for architects. *Architects' Journal*, pp. 82–83.

Becker, F. (1988a). Saunas and beer: Good multinational business. *Leaders, 11*(2), 82–83.

Becker, F. (1988b, February). Form follows process at dynamic Lloyd's of London. *Facility Design and Management*, pp. 54–58.

Becker, F. (1988c, January). Managing facility management. *Premises Management*, pp. 18–19.

Becker, F. (1988d). *The changing facilities organization*. Haverhill, England: PROJECT.

Becker, F. (1988e, June). The FM performance profile. *Premises Management*, pp. 17–18.

Becker, F. (1988f). Managing innovation: Computer and organizational ecology. In M. Helander (Ed.), *Handbook of human-computer interaction*. Amsterdam: Elsevier.

Becker, F. (in press). The total workplace: Facilities management and the elastic organization. New York: Van Nostrand Reinhold.

Becker, F., & Hoogesteger, A. (1986). Employee adjustment to an office relocation. *Human Ecology Forum, 15*(4), 6–9.

Becker, F., & Spitznagel, J. (1986). *Managing multi-national facilities*. Report to International Facility Management Association, Houston, Texas.

Becker, F., & Steele, F. (1990). The total workplace. *Facilities, 8*(3), 9–14.

Becker, F., & Syme, J. (1987). *Managing an office relocation*. Report to International Facility Management Association, Houston, Texas.

Berkowitz, L. (1954). Group standards, cohesiveness, and productivity. *Human Relations, 7*, 509–519.

Brill, M. (1984). *Using office design to increase productivity* (Vol. 1). Buffalo, NY: Workplace Design and Productivity, Inc.

Briscoe, G. (1983). Health and safety aspects of office automation: The European scene. *Proceedings of the world congress on the human aspects of automation*. Ann Arbor, MI: University of Michigan.

Brooks, M. J., & Kaplan, A. (1972). The office environment: space planning and effective behavior. *Human Factors, 14*, 373–391.

Caker, A., Hart, D. J., & Stewart, T. F. M., (1980). *Visual display terminals: A manual covering ergonomics, workplace designs, task organization, health and safety*. New York: Wiley.

Campbell, D. E. (1979). Interior office design and visitor response. *Journal of Applied Psychology, 64*, 648–653.

Clearwater, Y. (1980). *Comparison of effects of open and closed office design on job satisfaction and productivity*. Unpublished doctoral dissertation, University of California, Davis.

Coates, G. J. (Eds.). (1975). *Alternative learning environments*. Stroudsburg, PA: Dowden, Hutchinson & Ross.

Cohen, R., Goodnight, J. A., Poag, C. K., Cohen, S., Nichol, G. T., & Worley, P. (1986). Easing the transition to kindergarten: The affective and cognitive effects of different spatial familiarization experiences. *Environment and Behavior, 18*(3), 330–345.

Cooper, C. C. (1975). *Easter Hill Village*. New York: Free Press.

Dainoff, M. J., & Happ, A. (1981). Visual fatigue and occupational stress in VDT operators. *Human Factors, 23*(4), 421–438.

Davis, G., Becker, F., Duffy, F., & Sims, W. (1985). *ORBIT-2: Organizations, buildings, and information technology*. Norwalk, CT: The Harbinger Group.

DEGW. (January 1988). *Three cities compared: London, Tokyo, New York.* Report to Rosehaugh Stanhope Developers, London.

DEGW. (December 1987). *The space requirements of professional firms in the City of London.* Report to Rosehaugh Stanhope Developers, London.

DIO International Research Team. (1983). A contingency model of participative decision making: An analysis of 56 decisions in three Dutch organizations. *Journal of Occupational Psychology, 56,* 1–18.

Dolden, M., & Ward, R. (Eds.). (1986). *The impact of the work environment on productivity.* Washington, DC: Architectural Research Centers Consortium.

Duffy, F., Cave, C., & Worthington, J. (Eds.). (1976). *Planning office space.* London: Architectural Press.

Ekuan, S. (1982). *From office culture to office environment: A comparative study.* Tokyo, Japan: GK Industries.

Ellis, P., Becker, F., & Jockusch, P. (1987). *Achieving office quality: A report of international research on the nature of office quality and its relation to the processes of development, design, procurement, and management.* London, England: Building Use Studies.

Festinger, L., Schachter, S., & Back, K. (1950). *Social pressures in informal groups: A study of human factors in housing.* New York: Harper.

Francescato, G. (1979). *Residents' satisfaction in HUD-assisted housing: Design and management factors.* Washington, DC: US Department of Housing and Urban Development, US Government Printing Office.

Fried, M., & Gleicher, P. (1961). Some sources of residential satisfaction in an urban "slum." *Journal of the American Institute of Planners, 28,* 305–315.

Frogatt, C. (1985). *Analysis of innovative space and furniture allocation policies at the Union Carbide Corporation.* Unpublished master's thesis, Cornell University, Ithaca, New York.

Gans, H. (1961). Planning and social life: Friendship and neighbor relations in suburban communities. *Journal of the American Institute of Planners, 27*(2), 134–140.

Gans, H. J. (1968). *People and plans.* New York: Basic Books.

Glass, D., & Singer, J. (1972). *Urban stress: Experiments on noise and social stressors.* New York: Academic Press.

Harris, L., & Associates. (1981). *The Steelcase national study of office environments: Comfort and productivity in the office of the 80s.* Grand Rapids, MI: Steelcase, Inc.

Harris, L., & Associates. (1987). *The office environment index.* Grand Rapids, MI: Steelcase, Inc.

Hare, P., & Bales, R. (1963). Seating position and small-group interaction. *Sociometry, 20,* 480–486.

Hedge, A. (1989). Environmental conditions and health in offices. *International Review of Ergonomics, 3,* 87–110.

Homan, G. (1950). *The human group.* New York: Harcourt, Brace and World.

Justa, F. C., & Golan, M. B. (1977). Office design: Is privacy still a problem? *Journal of Architectural Research, 6*(2), 5–12.

Leduc, N. F. (1979, September). Communicating through computers: Impact in a small business group. *Telecommunications Policy,* pp. 235–244.

Lynch, K. (1972). *What time is this place?* Cambridge, MA: MIT Press.

Makower, J. (1981). *Office hazards: How your job can make you sick.* Washington, DC: Tilden Press.

Maslow, A., & Mintz, M. (1956). Effects of aesthetic surroundings: Initial effects of three aesthetic conditions upon perceiving "energy" and "well-being" in faces. *Journal of Psychology, 41,* 247–254.

McGregor, D. (1960). *The human side of enterprise*. New York: McGraw-Hill.

Michelson, W. (1977). *Environmental choice, human behavior, and residential satisfaction*. New York: Oxford University Press.

Morrow, P. C., & McElroy, J. C. (1981). Interior office design and visitor response: A constructive replication. *Journal of Applied Psychology, 66*(5), 646–650.

Naisbett, J. (1982). *Megatrends: Ten new directions transforming our lives*. New York: Warner Books.

Newman, O. (1975). *Design guidelines for creating defensible space*. Washington, DC: National Institute of Law Enforcement and Criminal Justice.

Oldham, G. R., & Brass, D. J. (1979). Employee reactions to an open-plan office: A naturally-occurring experiment. *Administrative Science Quarterly, 24*, 267–284.

Rice, A. K. (1958). *Productivity and Social Organization: The Ahmedabad experiment*. London: Tanistock.

Roethlisberger, F. J., & Dickson, W. J. (1939). *Management and the worker*, Cambridge, MA: Harvard University Press.

Sauser, W., Arauz, C., & Chambers, R. (1978). Exploring the relationship between level of office noise and salary recommendations: a preliminary research note. *Journal of Management, 4*, 57–63.

Schachter, S., Ellertson, N., McBride, D., & Gregory, D. (1951). An experimental test of cohesiveness and productivity. *Human Relations, 4*, 229–238.

Sherrod, D. (1974). Crowding, perceived control, and behavioral aftereffects. *Journal of Applied Social Psychology, 4*, 171–186.

Sommer, R. (1961). Leadership and group geography. *Sociometry, 24*, 99–110.

Sommer, R. (1967). Small group ecology. *Psychological Bulletin, 67*, 145–152.

Sommer, R. (1969). *Personal space*. Englewood Cliffs, NJ: Prentice-Hall.

Sommer, R. (1972). *Design awareness*. San Francisco: Rinehart Press.

Sommer, R. (1974). *Tight spaces: Hard architecture and how to humanize it*. Englewood Cliffs, NJ: Prentice-Hall.

Sommer, R. (1983). *Social design*. Englewood Cliffs, NJ: Prentice-Hall.

Sommer, R. (1986). The measurement of productivity. In M. Dolan and R. Ward (Eds.), *The impact of the work environment on productivity*. Washington, DC: Architectural Research Center's Consortium, American Institute of Architects.

Steele, F. (1973). *Physical settings and organizational development*. Reading, MA: Addison-Wesley.

Steele, F. (1977). *The feel of the work place: Understanding and improving organizational climate*. Reading, MA: Addison-Wesley.

Steele, F. (1986). *Making and managing high quality work places*. New York: Columbia University Press.

Steers, S. (1976). When is an organization effective? *Organizational Dynamics, 5*, 50–63.

Steinzor, B. (1950). The spatial factor in face-to-face discussion groups. *Journal of Abnormal and Social Psychology, 45*, 552–555.

Strodbeck, F. L., & Hook, L. H. (1961). The social dimensions of a twelve-man jury table. *Sociometry, 24*, 397–415.

Sundstrom, E. (1986). *Work places: The psychology of the physical environment in offices and factories*. New York: Cambridge University Press.

Sundstrom, E., Burt, R., & Kamp, D. (1980). Privacy at work: Architectural correlates of job satisfaction and job performance. *Academy of Management Journal, 23*, 101–117.

Taylor, F. (1911). *Principles of scientific management*. New York: Harper & Row.

Techau, D. A. (1987). *Organizational culture and the spatial ecology of areas*. Unpublished masters thesis, Department of Design and environmental analysis, New York State College of Human Ecology, Cornell University, Ithaca, New York.

Viteles, M. S., & Smith, K. R. (1941). *A psychological and physiological study of the accuracy, variability, and volume of work of young men in hot spaces with different noise levels.* (ASHVE Report). Washington, DC: Bureau of Ships, U.S. Navy.

Walton, R. E., & Vittori, W. (1983). New information technology: Organizational problem or opportunity? *Office Technology and People, 1,* 249–273.

Wandersman, A. (1979). User participation: A study of types of participation, effects, mediators, and individual differences. *Environment and Behavior, 11,* 465–482.

Weick, K. (1978). *The social psychology of organizing* (2nd ed.). Reading, MA: Addison-Wesley.

Whyte, W. F. (1948). *Human relations in the restaurant industry.* New York: McGraw-Hill.

Working Women (1981). *Warning: Health hazards for office workers—An overview of problems and solutions in occupational health in the office.* Cleveland: Working Women Education Fund.

Worthington, J., & Konya, A. (1988). *Fitting out the workplace.* London: Butterworth.

Zeisel, J. (1974). Fundamental values in planning with the nonpaying client. In J. Lang, C. Burnette, W. Moleski, & D. Vachon (Eds.), *Designing for human behavior* (pp. 293–301). Stroudsburg, PA: Dowden, Hutchinson & Ross.

Zube, E. H., & Sell, J. L. (1986). Human dimensions of environmental change. *Journal of Planning Literature, 1*(2), 162–176.

Zweigenhaft, R. L. (1976). Personal space in the faculty office: Desk placement and the student–faculty interaction. *Journal of Applied Psychology, 64,* 529–532.

5

American Vernacular Architecture

THOMAS C. HUBKA

A chapter for an "advances" volume could begin with a brief outline of the subject before examining the leading, cutting-edge works. The scope of American vernacular architectural studies, however, needs considerable explanation, because the range of topics now considered "vernacular" has changed in the last 10 years, causing much restructuring in the field. The idea of a succinct or unified subject has been strained by the inclusion of large unwieldy new areas such as multiunit and mass-produced housing, commercial and industrial architecture, and urban structures of all periods and types (see Figures 1–3). These new areas promise to challenge the central themes and the existing agenda within this loosely organized discipline. Consequently, defining the scope of American vernacular architecture studies takes on greater importance for developing an idea of "advances" for this domain of inquiry.

DEFINING VERNACULAR ARCHITECTURE

In retrospect, the expansion of present American vernacular architecture studies has followed an inevitable trajectory as the outline of the field has extended outward to its ultimate material boundaries—the

Thomas C. Hubka • Department of Architecture, University of Wisconsin–Milwaukee, Milwaukee, Wisconsin 53201.

FIGURE 1. Brownstone apartments (1892–1893), Hamilton Heights, Brooklyn, New York. From Diane Maddex (Ed.), *Built in the USA*, Washington, DC: Preservation Press, 1985. Reprinted by permission.

FIGURE 2. Marine A grain elevator (1925), Buffalo. From Reyner Banham, *A Concrete Atlantis*, Cambridge, MA: MIT Press, 1986. Reprinted by permission.

FIGURE 3. Commercial buildings, Michigan Boulevard, Chicago. From Chester H. Liebs, *Main Street to Miracle Mile*, Boston: Little, Brown, 1985. Copyright 1985 by Little, Brown, Inc. Reprinted by permission.

entire built environment—everything except, perhaps, academic or high-style architecture. The shock to people outside American vernacular architecture studies is the ambitious scope. The good news is that the previous ambiguity about what constituted the scope of vernacular architecture studies has finally been clarified. The bad news is that it now includes every building ever built (even those continuing to be built), at any time, any place, by any people, and for any purpose— every last one! This awesome realization (while not shared uniformly) has already begun to change the perception and direction of vernacular architecture study. Many vernacular scholars have welcomed the territorial expansion as a sign of continuing vitality and increased interest in the field, and few will miss the arcane discussions about what buildings really constitute a vernacular architecture.

But this expansion has also generated problems of clarity and direction. The idea of understanding or categorizing the content of vernacular architecture has been complicated by the inclusion of such odd bedfellows as: Civil War temporary camp shelters (Nelson, 1982), highway "strip" architecture (Liebs, 1985), ethnic front yard sculptures (Sciorpa, 1989), early garages (Goat, 1989), and concrete grain elevators (Banham, 1986).

Yet the quantitative problems pale beside the qualitative. Along with territorial expansion has come new interdisciplinary perspectives for analyzing vernacular architecture. Even the common English barn (in America) may be interpreted from material, historical, semiotic, social–psychological, anthropological, or cultural symbol perspectives, often with radically different results; compare, for example, English barns as analyzed in Hart (1975), St. George (1986), Glassie (1974), Hubka (1984), and Noble (1984). If response to growth, therefore, was merely a problem of quantitative restructuring, it might be more easily solved, but competing interpretations from different disciplinary and theoretical perspectives have caused vernacular scholars to reconsider some of their basic suppositions.

The present status of vernacular studies may be compared to a Darwinian period of intense biological curiosity when the entire natural universe was suddenly thrown open for exploration (Himmelfarb, 1959). Exhilarating scientific ferment, fueled by the exploration of new continents, made the prospect of discovering, naming, and grouping all earthly plant and animal species a sudden reality. The new discoveries were electrifying, but the real excitement came later when researchers such as Darwin began to question what the new species meant in relation to the old species. The results were startling, finally turning the old biological world upside down. Perhaps vernacular architecture study is approaching that exhilarating Darwinian plateau, as the outline of the entire architectural world begins to emerge. Like Darwin's contemporaries, we should prepare ourselves for some shocks of reorganization and reinterpretation. But this is jumping ahead—first, vernacular basics.

It has become a standard convention for interpreters and catalogers of vernacular architecture to ritually chastise, denounce, apologize for, and finally grudgingly accept under protest the term "vernacular architecture" (Lawrence, 1983; Riley, 1987; Upton, 1983; Upton & Vlach, 1986). This ritual slaughter and reincarnation might appear to be an amusing literary convention, but like seemingly meaningless ritual acts, it conceals a deeper totemic logic. Encapsulated within the word *vernacular* lies its ancient, latent meaning, which continues to bedevil its adherents. In Latin, *vernaculus* literally means a native slave born in a master's house. What polite dictionaries conceal is the ideological prejudice, the pejorative sting: uncouth, from the provinces, provincial, hick, hayseed, yahoo, and so on. I hasten to emphasize that today the word vernacular has largely lost this historical pejorative connotation; for some writers the term might even be considered avant-garde, as for example in Hirshorn and Izenour (1973) and Venturi, Brown, and Izenour (1977). Nevertheless, to appreciate the current status of vernacular scholarship, it is necessary to appreciate the lingering, subliminal strength of this historic distinction—between vernacular architecture and, by omission, other architecture.

Vernacular and Elite

The study of vernacular architecture began outside traditional academic architectural studies, and this fact has significantly influenced its growth. Beginning in the early nineteenth century, vernacular architecture study gradually emerged from historical culture and ethnographic studies, fueled by European and American nationalistic and romantic revivals (Upton, 1983). During the twentieth century, vernacular buildings slowly attracted a wider, multidisciplinary following, principally since the 1960s. Today it exists as a topic of study in a variety of humanistic disciplines such as English, folklore, archaeology, anthropology, geography, history, American studies, and the environmental and design professions including architecture, landscape architecture, art history, and environmental and behavior studies (Upton, 1983; Wells, 1986).

Vernacular architecture study developed independently from the well-established academic discipline of architecture, but its powerful traditions have influenced the way vernacular architecture has been perceived. Traditional academic studies have generally analyzed structures of the upper or ruling classes, described alternatively as monumental, avant-garde, aesthetic, or elite architecture. The academic study of architecture emerged during the European Enlightenment as a component of classical and humanistic studies and was reformulated into a separate discipline related to art history during the realignment of academic disciplines in the second half of the nineteenth century (Allsopp, 1970). Because vernacular structures were seldom explored by architectural historians, whose considerable accomplishments were lavished on a narrow band of the architectural spectrum, a vast body of American architecture remained unexplored until quite recently, It is into this scholarly vacuum that the present horde of multidiscipline explorers have converged.

Today, rigid distinctions between elite and vernacular topics have blurred considerably, but the underlying issues that generated these categories have lost little of their original meaning. Formalized for the Romans by Vitruvius (1899/1960) and given forceful expression by John Ruskin (1884) in the nineteenth century, the distinction between "architecture" and "building" entered the modern era as a division between beautiful and monumental architecture on the one hand and functional and utilitarian buildings on the other, as for example in Pevsner (1976) and Fletcher (1956). Although often portrayed as an aesthetic or moral issue (Scruton, 1979), this distinction is essentially an economic and social division for ranking buildings according to prevailing cultural taste or ruling logics (Watkin, 1977). It is important to recognize the continuing existence of this ancient distinction between elite and vernacular, especially in professional architecture schools, as for example in Ivy (1989), but also subliminally in popular culture (Gans, 1974).

For many of the people who examine vernacular architecture today, the distinction between high and low architecture has faded into a tangled middle ground—popular architecture. The term "popular" was codified for vernacular historians in 1968 by Henry Glassie in *Patterns in the Material Folk Culture of the Eastern United States.* Today it is the most commonly accepted term for describing the architecture existing between the poles of high-style, elite, or architect-designed architecture and vernacular, folk, traditional, or preindustrial architecture (see Figures 4–6). The term "popular" was necessitated by the inability of either vernacular or elite categories to adequately account for the massive body of late nineteenth- and twentieth-century architecture, especially buildings that were the product of a modern, industrialized society. Pressure to include these buildings caused the definition of vernacular to shed its traditional association with rural, preindustrial buildings and to include an expanded range of nineteenth- and twentieth-century architecture. Today the terms *popular* and *vernacular,* or even *popular-vernacular,* are loosely interchangeable with *vernacular,* usually referring to older, less numerous, indigenous, or rural buildings and *popular* referring to newer, more numerous, or urban buildings. The tripartite distinction between *high-style, popular,* and *vernacular* is the currently accepted, if not in all ways satisfying model for organizing the total volume of American and worldwide architectural production (Glassie, 1968).

The current expansion of both vernacular and elite architectural studies to address popular architectural topics has been accomplished by simultaneous movements at both ends of the architectural spectrum. In the past 10 years, architects and academic architectural historians have significantly enlarged their discipline to include a larger portion of the popular spectrum. For example, the Society of Architectural Historians, long a bastion of insulated, high-style architectural studies, now

FIGURE 4. Woolworth Building (1913), New York. From Paul Goldberger, *The Skyscraper*, New York: Alfred A. Knopf, 1986. Reprinted by permission.

FIGURE 5. Safeway supermarket (c. 1980), Laramie, Wyoming. From Chester H. Liebs, *Main Street to Miracle Mile*, Boston: Little, Brown, 1985. Copyright 1985 by Little, Brown, Inc. Reprinted by permission.

FIGURE 6. Lesser Dabney house, Louisa County, Virginia. From Henry Glassie, *Folk Housing in Middle Virginia*, Knoxville: University of Tennessee Press, 1975. Reprinted by permission.

includes a wide range of popular and vernacular paper sessions at its annual meetings, as for example in Pearson (1989). From the opposite end, vernacular architectural studies, long a bastion of rural, preindustrial, domestic building studies now include most of the popular architecture spectrum, even high-style architectural topics (Lounsbury, 1989). The result of these simultaneous movements has produced a significant democratization of architectural studies and the prospect of a rare comprehensiveness for the entire architectural field.

VERNACULAR PAST AND PRESENT

The loose discipline of vernacular architectural studies is old enough to have produced competing interpretations of its subject matter. Among the many subgroups that explore vernacular topics, a significant division has occurred between those that explore historic vernacular architecture, the "pasts," and those that explore current or existing vernacular architecture, the "presents." On one side, the "pasts" are aligned with the humanistic disciplines and advocate historical-humanistic approaches to the study of vernacular architecture as typified by Cummings (1979) and Upton (1986). On the other side, the "presents" are aligned with the social and scientific disciplines and advocate the study of present, ongoing, or living vernacular architecture as typified by Kowinski (1985) and Saile (1985). Both groups have produced different types of studies with different consequences for vernacular architecture research. This author is affiliated with the "pasts" who worry about "present" vernacular studies lacking sufficient grounding in the past, but "presents" counter with concerns about the scientific method of the "pasts" and their ability to grasp current models of social–cultural research.

It seems obvious that there are important points to be gained from both sides. A wide spectrum of environment-behavior and anthropological–architectural studies of ongoing vernacular architecture have made important contributions to the study of vernacular environments (Cooper, 1975; Rapoport, 1983a; Zelinsky, 1973). Obviously all vernacular study need not grapple with the past. If current anthropology or social research is the goal, then architecture may certainly be another vehicle for study. In Third World studies, where the relationship between historical architecture and present architecture might be much more fluid and ongoing, vernacular studies need not emphasize the analysis of the ancient past as, for example, in Ustunkok (1988). In such cases, studies of the evolving present are much more germane.

Vernacularists who study the history of architecture caution, however, that sooner or later the need for a comprehensive understanding of how past architecture becomes present will arise, and it must be answered by the finest studies of the historical kind (Glassie, 1975). On the other hand, vernacular historicists have much to learn from current vernacular studies concerning the relationship between people and their environments, certainly one of the weakest links in historical vernacular scholarship, as acknowledged in Upton (1983). From either perspective, the best vernacular architecture scholarship rests on a comprehensive understanding of a building in its cultural context, and this remains true for works of the past and the present.

The problem of defining advances for vernacular architecture study within this multisided context is thus a complicated one. Sudden expansion into the full range of architectural production has produced a rich diversity of topics, but will the center hold? Is there a core of studies, and will it sustain long-range development and a future of advances in vernacular architecture?

I feel the present would-be discipline of vernacular studies is experiencing a youthful crisis of identity—a coming to terms with what it is that is vernacular. In order to clarify the issue, it is my hope that the term *vernacular architecture* will someday be viewed as a transition vehicle, a historical catalyst made necessary for gathering followers, focusing efforts, and gaining academic legitimacy. It has helped unite a diverse group of scholars under a commonly perceived banner of appreciation for common buildings. Having achieved a small measure of academic respectability, the once second-class "vernacular" citizen should no longer find the term necessary. I argue for the gradual abandonment of the term principally because its historic pejorative connotation has begun to inhibit creative efforts to integrate "vernacular" buildings into the total body of architecture. Hopefully, vernacular study will become a more integrated part of architectural study with lots of interesting subcategories, and especially without a capital "A" and a small "v." One way to gradually resolve this "vernacular" problem is to reinterpret the distinction between *elite, popular,* and *vernacular* architecture.

I argue that the tripartite elite–popular–vernacular classification model is improved by a slightly modified version emphasizing the distinction between traditional and progressive ideas, processes, and buildings that constitute the architectural totality. This dialectical strategy recognizes a fundamental distinction between old and new, permanence and change, and stasis and flux at every level of the architectural ensemble including conceptualization, design, history, materials, function, culture, and context. Such a restructuring between traditional and progressive factors for all architecture would acknowledge the real strengths of the current elite–popular–vernacular distinction, including differences between design process and stylistic characteristics, while democratizing categorization by eliminating the cultural and class distinctions currently accepted. Such a revision, I feel, would also facilitate the long overdue analysis of buildings according to multiple, comprehensive criteria, including both traditional (permanent) and progressive (changing) characteristics. This approach to architectural analysis would recognize pluralistic criteria for building analysis, not only (but not excluding) the traditional art history criterion of architectural style. For example, a house with a traditional structural system and progressive stylistic system is perhaps the popular architectural norm (Figure 7). Its windows might be organized in a traditional arrangement but employ a new technology, and yet its inhabitants might use the windows creatively in order to accommodate an activity of long standing. It is this type of comprehensive packaging of multiple criteria for architectural analysis that has long been advocated by such diverse architectural theorists as William Kleinsasser (1979) and Amos Rapoport (1983a, 1983b) and could be used to achieve a more architecturally comprehensive and culturally responsible analysis of the architectural product and its physical and cultural context set within a historic continuum.

I would, therefore, recommend the gradual abandonment of the vernacular "v" word and the elite "e" word and their replacement with terms such as "traditional" and "progressive" referring to all buildings.

FIGURE 7. Houses on Stockton Row (1880s), Cape May, New Jersey. From Diane Maddex (Ed.), *Built in the USA*, Washington, DC: Preservation Press, 1985. Reprinted by permission.

As the defensive need for the term "vernacular" evolves into a healthier concept of "traditional" (or popular, or folk), a more comprehensive, pluralistic attitude could and should emerge. Such an attitude would recognize the study of architecture as a highly integrated cultural continuum necessitating a subdivision of the whole into manageable but interrelated parts. Although these new subdivisions will invariably be influenced by the unequal whims of current scholastic (and more broadly, ideological) fashion, the new order must be made responsible for forging a comprehensive account of the whole body of architecture (Hubka, 1985a).

The term *vernacular architecture*, born of elitism and historic neglect, is presently benign, but it remains an inhibiting term, a plantation term whose time has run out. Hopefully it will vaporize into a comprehensive architectural continuum based on the ideas of cultural plurality and consensus; twin goals for generating the unbiased account of what happened to all buildings. These changes, while substantial, need not produce a much feared populist reign of terror (Trachenberg, 1988–1989), including the destruction of aesthetics (Scruton, 1979), but should allow the story of vernacular architecture to be told alongside the story of all architecture, as begun by architectural historians such as Kostof (1985) and Roth (1979).

Before relinquishing the term, however, vernacular advocates will be cautious and watchful, lest the liberated (vernacular) slave be returned to the master's (high-style) house (Hubka, 1985b). Remembering their humble origins, they will insist that the new architectural order not perpetuate the old order of economic, class, and cultural elitism, but create an order which accounts for all buildings in a hard-fought consensus about what actually happened. It is a difficult goal, but it is vernacularly worth the struggle.

IDEOLOGY: THE HIDDEN AGENDA

All disciplines, even fledgling ones like vernacular/traditional building studies, can be identified by a body of shared precepts, assumptions, or common beliefs that constitute an ideological framework or, simply, its point(s) of view. While hardly ever uniform or monolithic, these unquestioned assumptions structure a hidden agenda sanctioning certain types of inquiry, limiting others (Foucault, 1980; Geertz, 1973). Often these quasi-religious beliefs are unquestioned by their practitioners, and only in retrospect do the limitations of a particular ideological framework or set of assumptions become evident.

SILENT ARTIFACTS AND COMMON PEOPLE: THE "NEW HISTORY"

Some of the most significant recent work of vernacular/traditional architecture scholarship investigates buildings where little was known about their builders or the people who lived in them, as for example in Jordan (1985; see Figure 8). Often these buildings were made by common people outside the mainstream of traditional historical investigation. The recent focus upon unknown common people and their buildings is, of course, not without precedent and should be seen as the effects of a broad popular democratic movement occurring throughout the humanistic disciplines. The phenomena has been described by many as the "New History," or the history of common people (Glassie, 1975). Although the origins and dates may be debated, the results have had a profound effect on the humanistic disciplines. Within the last 30 years the center of humanistic studies has shifted from the well-documented history of the few (usually the wealthy, literate, powerful, and known) to the less well documented history of the many (usually the poor, illiterate, unprivileged, and unknown). While the effects and wisdom of this reorientation will continue to be debated (Bloom, 1987), the complete triumph of the New History model for vernacular studies cannot

FIGURE 8. Saddlebag House, Louisa County, Virginia. From Henry Glassie, *Folk Housing in Middle Virginia*, Knoxville: University of Tennessee Press, 1975. Reprinted by permission.

be overemphasized. In fact, the primary subject matter for vernacular/traditional architecture studies constitutes a program for topics of common people and their buildings. (It might even be argued that the emergence of New History precipitated the widespread scholarly interest in vernacular/traditional architecture.)

Anonymous vernacular/traditional buildings have, of course, always been available to study. What has changed is the will to study them (spurred by New History), and the availability of more sophisticated methods of study; consequently, what can be studied has increased. Lead by Henry Glassie's *Folk Housing in Middle Virginia* (1975), vernacular architecture historians in America have applied a range of interdisciplinary theory and methods, including anthropological, semiotic, and material cultural strategies, to the study of anonymous buildings once considered unapproachable for lack of literary documentation. The result has been a quickened expansion of the field as researchers have rushed to analyze buildings long considered unproductive or impossible to interpret.

VERNACULAR EXCEPTIONALISM: THE NEW ROMANTICISM

The ideology of New History is not without problems for vernacular/traditional architecture studies. One result has been the establishment of a reverse elitism in which vernacular/traditional subjects are perceived as exceptional and privileged simply because they are anonymous or folk—a phenomenon the architectural historian Dell Upton has labeled "vernacular exceptionalism" (1988; see Figure 9). It is an amusing problem for veteran vernacular researchers long accustomed to minority status in a world previously dominated by an academic architecture establishment. Nevertheless, academic maturity combined with a popular appreciation for vernacular architecture has created an intellectual climate where preindustrial building and old-time ways have occasionally been seen as better, purer, and more decent than other forms of architecture, particularly modern architecture, a theme begun in Sloane (1967) and Rudofsky (1977) and continued in works such as Sadler & Sadler (1981). A continuation of nineteenth-century romantic anti-industrialism, this attitude is one outgrowth of the success of vernacular study. Having gained a measure of academic maturity, vernacular/traditional architecture scholarship need not isolate and defend its subject matter with exceptionalist claims. I am arguing here for the abandonment of vernacular romanticism and exceptionalism in all forms and for the need to integrate vernacular studies into a larger body of architectural and cultural scholarship.

FIGURE 9. Tobacco barn, Louisa County, Virginia. From Henry Glassie, *Folk Housing in Middle Virginia*, Knoxville: University of Tennessee Press, 1975. Reprinted by permission.

PARADIGMS: THE ACKNOWLEDGED AGENDA

Unlike unquestioned ideological influences, paradigms guide research exploration on a working plane of disciplinary discourse. They are analyzed and deployed with a degree of conscious recognition and are subject to both the whims of academic fashion and the most profound developments of intellectual achievement.

I have listed four frequently employed investigative paradigms, or models, that I feel have guided the most significant works of vernacular/traditional architecture scholarship during the last 10 years. Unlike other recent summaries of vernacular scholarship organized by discipline (Marshall, 1981), methodological approaches (Upton, 1983), content (Riley, 1987), and definitions (Lawrence, 1983), the following list cuts diagonally through several of those groupings in order to reveal a loose bundle of strategies that seem to unite current studies of vernacular/traditional buildings.

ARTIFACT AND MEANING: WHAT IT IS AND WHAT IT MEANS

Perhaps the most significant development in vernacular scholarship during the last 25 years has been the gradual emergence and near dominance of artifact–meaning-oriented research replacing nineteenth-century artifact- or object-oriented research as the primary focus of vernacular/traditional architecture investigation (Upton, 1983). For vernacular scholars this has meant a reformulation of the descriptive bricks-and-boards type of question, "what was it?" to include the cognitive-symbolic question, "what did it mean (to the people who built and inhabited it)?" Although architectural scholarship has always interpreted meaning, it is the active search for the cognitive-symbolic component of building that particularly marks this new orientation to scholarship. Dell Upton has described three approaches to cognitive-oriented vernacular research: social, cultural, and symbolically oriented studies (Upton, 1983). Each approach is grounded in the assumption that artifacts such as vernacular architecture can be probed to reveal mental structure (whether socially, culturally, or symbolically based) about the people and culture that made and inhabited them. Of the three approaches, the search for symbolic meaning, guided by linguistic and semiotic models, has proven the most fruitful during the last 10 years (Glassie, 1975; Jencks & Baird, 1970).

Current efforts to interpret the symbolic character of vernacular/traditional architecture must confront the bewildering problem of understanding the meaning of an individual building within a larger popular cultural context. This task is difficult enough for a specific building designed by a specific architect (for example, a structure by Andrea Palladio or James Sterling) where considerable information is available about the designer, owner, builder, and the building's cultural context. The task is exceedingly more complex when confronting the problem of assigning meaning to a typical vernacular/traditional structure, where, for example, the architect and owner might be unknown and the user or client may or may not be known and may or may not have taken part in the design or building process. Even when substantive information is available, the vernacular scholar must establish the cognitive linkage between different and often unrelated participants (designer, owners, builders, users) and establish the complex patterns of meaning between the building and its larger cultural context. These are the front lines of cognitive–symbolic-oriented research for vernacular/traditional architectural studies, of which it cannot be said that anyone has yet hit upon the ideal methodology.

When considering the multiplicity of vernacular building types, the plurality of owners, builders, and users, and the wide-ranging effects of a commercial–industrial economy, it seems unlikely that a single, unified theory will emerge. Still, a long-term goal of all future artifact-meaning studies must be to locate cognitive-symbolic structures closest to the time, place, and people that made or inhabited particular vernacular structures. It is always easier to interpret the meaning or symbolic content from a later period or in relation to contemporary values. To account for the day-to-day meaning for the people who actually made or originally inhabited these structures, however, is the ultimate task for object–meaning-oriented studies—we have a considerable distance yet to travel. Perhaps the strongest work to date is Glassie's *Passing the Time in Ballymenone* (1982).

BUILDING DOMINANCE AND THEORY SUBORDINANCE: ARTIFACT POSITIVISM

One reason why vernacular/traditional architecture scholarship is different from traditional academic scholarship is that vernacular architecture researchers never seem to have lost their tactile infatuation with the buildings they study. Perhaps they never relinquish their pleasure in climbing around old buildings and searching for mysterious clues in dusty attics. Whatever the reasons, a deep-rooted materialistic faith seems to unite scholars in a profound trust of the physical artifact—a faith in bricks and boards as the final testing place for analytical theory (Glassie, 1977).

To those familiar with the general development of architectural scholarship, this assessment might seem hopelessly quaint or overstating the obvious; certainly all architectural scholarship must be based upon the physical evidence of buildings. Yet it is the degree to which vernacular scholars finally rely upon the physical evidence of particular buildings (rather than generalities about buildings) to form their theories that is the particular mark of the best vernacular scholarship. There are several reasons for this positivistic artifact approach.

Vernacular scholarship originated outside traditional architectural scholarship and never developed a neoclassicizing, neoplatonic theoretical tradition for interpreting architecture as exemplified, for example, by Rykwert (1981) and Scruton (1979). For initial vernacular scholars there was no Vitruvius or Piranesi, and certainly no Boullee to insert idealistic, platonic prescriptions between the researcher and the vernacular building. I would quickly emphasize that vernacular studies have not been characterized by a naive, materialistic positivism nor as a group that has ignored theoretical approaches to building analysis. In fact, lack of historical literary documentary evidence has necessitated creative theorization at all levels of vernacular study (Glassie, 1975; Upton, 1983).

Analytical constructs for vernacularists are, however, placed in subordination to material evidence in a way unprecedented in traditional high-style architectural analysis. The bottom line for vernacular scholars is simply that the physical evidence of buildings is trusted more than theories about buildings. Architectural evidence may stand in various relationships to theoretical analysis, but in the final analysis vernacular scholars seem to elevate the material evidence of bricks and boards over conceptual ideals or suggestions of theory. It is this faith in the primacy of the material artifact that has continually been the mark of the best vernacular scholarship and will probably continue to be so in the future (Bronner, 1983, 1985; Glassie, 1977). I maintain that no substantive work of vernacular/traditional scholarship has been produced that did not have at its core the finest, substantive, material knowledge of building. Furthermore, this knowledge about buildings usually preceded, and was not preceded by, theoretical concepts. (This is certainly not the case in high-style architecture studies [Rykwert, 1981].)

PLURALISM OF CONTENT, THEORY, AND METHOD: A POPULIST STRATEGY

As the quantity of various vernacular/traditional building types has expanded, so has the number of scholars from various disciplines who explore them. Today the field contains researchers from most academic disciplines except, perhaps, the physical sciences. Significantly, no discipline or even a set of disciplines dominates the field. Disciplinary plurality could be attributed to many factors, including intellectual immaturity, but it is more likely indicative of a richness in content as well as a variety of problem topics posed by vernacular/traditional architecture research. It now seems obvious to many researchers that fundamentally different strategies might be employed when investigating an anonymous seventeenth-century Massachusetts salt-box dwelling and a contemporary Los Angeles shopping center. Yet even this extreme comparison could be shown to have substantive similarity if both investigations explored the same issues and asked the same types of questions—but they do not and probably should not. Vernacular/traditional architecture studies invite a pluralism of theory and method because the research aims at different types of issues where fundamentally different types of buildings pose fundamentally different types of research questions—what is known and unknown about one type of building might be fundamentally different from another building. A consequence of vernacular/traditional methodological plurality has produced a creative climate for research described by Henry Glassie as an intellectual and interdisciplinary poaching exercise—a kind of opportunist, carpetbagging, multi-discipline-hopping scholarship (Glassie, 1977). The spirit of this cross-disciplinary orientation accurately describes an obvious strength of recent vernacular scholarship—its multidisciplinary, multitheoretical intellectual tradition, unusually free of narrow, party-line, intradisciplinary theorization.

To examine the best work of vernacular/traditional architecture scholarship during the last 10 years is to review works characterized by a very broad approach to interdisciplinary study. Even works reputed to emphasize a particular research method, for example an analysis of the widely cited structuralist method of Henry Glassie in *Folk Housing in Middle Virginia* (Glassie, 1975), reveal several broad theoretic strategies, certainly not only structuralism. Future vernacular/traditional studies will probably never adhere to a unified theory or method. Methodological and theoretical pluralism should not, however, be misinterpreted as a sign of unchecked relativism, intellectual or moral nihilism, or a weakness in scientific method. In a loosely organized discipline where theory is subordinate to artifact, it is simply not important or possible to seek a unified theory. In a discipline dominated by a faith in the material artifact as the bottom line of all intellectual inquiry, theory will continue to be subordinate to the artifact in building.

Pattern in Building: Variations on a Theme

The ability to discern pattern—the forest from the trees—is a fundamental component of human thought (Gombrick, 1979; Langer, 1953) and a significant component to vernacular/traditional architecture scholarship (Ching, 1979; Hubka, 1985a; see Figures 10 and 11). While pattern seeking is employed in all architectural scholarship, it is not just another method for vernacular scholarship but a fundamental assumption about the nature of buildings that are often closely unified and repetitive. Unlike the study of a particular building, such as a specific *palazzo,* where the essential meaning may or may not be related to others of its kind, vernacular architecture gains meaning and stature primarily in relation to others of its kind. While a vernacular/traditional building can be analyzed as an individual object, it is in the collective that its meaning often achieves significance.

Once again, it has been Henry Glassie who has led the way in refining and broadening the idea of pattern in vernacular environments. By linking linguistic patterns of grammar transformation to explain variation on a theme, he has constructed a challenging explanation for the existence of common eighteenth- and nineteenth-century farm structures that has led the way for further research (Glassie, 1975).

The importance of pattern recognition for the vernacular/traditional building scholar is a surprising revelation to architects and architectural

FIGURE 10. Colonial cottage court (1930s) near Louisville, Kentucky. From Diane Maddex (Ed.), *Built in the USA*, Washington, DC: Preservation Press, 1985. Reprinted by permission.

FIGURE 11. Harrison Avenue (c. 1900), Guthrie, Oklahoma. From Diane Maddex (Ed.), *Built in the USA*, Washington, DC: Preservation Press, 1985. Reprinted by permission.

historians usually accustomed to searching for "original" structures or buildings demonstrating earliest characteristics of a type or style (Rykwert, 1981). Origins or "first buildings" for the vernacular historian are theoretically interesting but misleading, since single buildings are rarely significant to the development of most vernacular/traditional architectural types (Hubka, 1984). What vernacular scholars seek about the origins of a building is commonality and consensus—the typical (patterned) example of a kind. This marks an analytical gulf between elite and vernacular scholarship of the greatest possible consequences. What is often important about a particular vernacular/traditional building is not its originality or its uniqueness but its commonality—what it shares with others of its kind. The problem of origins and uniqueness is, of course, important to all architectural scholarship, including vernacular. The emphasis on origins, however, loses its platonic distinctiveness when it is recognized that the origins of vernacular/traditional buildings are multiple, pluralistic, fuzzy, and historically elusive. There is certainly no original platonic log cabin in the sky nor was there ever one on the ground—only previous and changing log cabins with rich histories stretching back in time. The center of vernacular/traditional building scholarship therefore decisively shifts from an origin-seeking, precursor-focusing, art history model, to a culturally comprehensive ideal of consensus sharing and collective development. The vernacular scholar finds the shifting center of normality, consensus, and shared values, from which explorations can be made to earlier precursors and later developments.

IMPLICATIONS FOR FUTURE RESEARCH AND APPLICATIONS

Today American vernacular/traditional architecture study is an expanding subject area (like the subject of vernacular landscapes [Riley, 1987]), but it is not an established academic discipline with a consensus of theory or method. It could be swallowed whole by a single discipline or carved up into several related disciplines. To those who study vernacular/traditional buildings, the issue raises little passion or concern and is indicative of a deeper commitment to the primacy of the subject matter, not the discipline. The buildings and cultures vernacular historians study are simply more important than allegiance to a particular methodological strategy or disciplinary agenda. As a general rule, the best new work about a vernacular/traditional building subject will lead the way. For example, if a major new work is produced about West Texas banks in the early nineteenth century, watch for the stock of commercial vernacular/traditional building studies to rise. In any event, the future direction for vernacular studies will probably be heavily influenced by the way popular architecture topics are interpreted and integrated into the context of previous studies. The following are other directions for vernacular studies.

Functional Categories: Organizing the Field

Now that the total architectural pie has been served up, it awaits the carving and labeling. I recommend categorization and classification primarily according to comprehensive functional criteria. For example, functional classification might unite all housing (high style and vernacular) and divide the rest of the building pie into manageable portions such as commercial, industrial, collegiate, and the like (see Figures 12 and 13). This type of functional categorization has existed for a long time (Pevsner, 1976; Vitruvius, 1889/1960), but it has not significantly structured the academic analysis of architecture. Quantitative organization according to function will force major revisions to established architectural theory. For example, Reyner Banham's (1986) perceptive examination of the midwestern grain elevator promises a fundamental re-

FIGURE 12. Train station (1914), Mountain Lake, Minnesota. From John R. Stilgoe, *Metropolitan Corridor*, New Haven: Yale University Press, 1983. Reprinted by permission.

examination of some of the basic tenets of modern architecture. Paul Turner's (1984) grouping of collegiate architecture according to function, not style, has forced a long overdue assessment of the importance of educational buildings in American architecture. Obviously, these books have only begun to scratch the surface of functional building reinterpretation.

DEMOGRAPHICS: NUMBERS COUNT

A significant potential for strengthening American vernacular/ traditional building studies exist in the development of comprehensive demography—the tabulation of everything that has ever been built. Such an accurate picture could upset the applecart of many insular, idiosyncratic, isolated studies. For example, how numerous and how significant were the legendary and overemphasized Garrison houses in New England, round barns in the North, and dogtrot houses in the South?

Initial efforts have been made to count, describe, and arrange vast regions of architecture by city and state historical societies and state historic preservation departments. While only a theoretical possibility until the 1960s, recent computer technology developments have suddenly made this populist dream a tempting possibility. But before cele-

FIGURE 13. Frame churches (1938), near Winner, South Dakota. From Diane Maddex (Ed.), *Built in the USA,* Washington, DC: Preservation Press, 1985. Reprinted by permission.

bration can begin, some sobering assessments are in order. As vernacular researchers who have worked with city or state historical preservation offices know, much of the existing vernacular/traditional building data is seriously flawed, especially with regard to the most common buildings (Wyatt, 1986). Future research relying on this type of survey information must be critically examined with regard to the reliability of the database. Before recommending the destruction of all previous data, I would like to be clear that some existing data will be helpful and precise enough for some studies (for example, when counting residences or assessing owner occupations). But if it is to be used to speculate about more complex, comprehensive architectural issues (such as room usage, the influence of progressive styles, or construction systems), it will be seriously flawed and there is little hope of correcting these flaws. (The old–new computer adage is still true: the research will be only as good as the original data.)

What to do? We will probably never again see a level of government funding for surveys as was made available to state preservation offices in the 1970s. Therefore, I suggest careful demographic case studies in sample, representative areas from which to speculate about larger and larger areas. I have argued elsewhere that one of the finest tests for comprehensive architectural scholarship is accurate demographic data. (Hubka, 1985a). The greatness of Monticello is seen more clearly within its vernacular context (Glassie, 1975). When we see the ratios and literally the numbers of buildings (for slaves, yeoman, and planters), we approach a comprehensive democratic ideal of historical totality—the record of all the buildings and the lives of all the people—simultaneously. This is, of course, often preached but rarely practiced. If practiced, it would finally nudge elite studies to proclaim their elitism.

What I am decidedly not advocating is a new emphasis on computer-guided number-crunching statistics. I am advocating that there should be an ideal of vernacular objectivity confirmed in demographics and that the counting of buildings provides one of the finest leveling methods for assessing the significance of all architectural research.

Popular Architecture: Between Elite and Vernacular

The historical relationship between vernacular/traditional architecture (usually preindustrial, rural) and popular architecture (usually late nineteenth and twentieth century, industrial, urban, and suburban) continues to erode, forcing reevaluations of the elite and vernacular extremes. The scope and diversity of popular architecture subjects has become so immense that it will invariably restructure the present outlines of all architectural study. Several important areas will have the greatest impact upon future advances.

Designer–client in mass culture. The traditional relationship between architect–builder–client must be radically reformulated in a context of mass culture, mass communications, and national economic and technological systems (Kowinski, 1985; Figure 14). New research that significantly interprets the role of the design process in mass culture will

Figure 14. Gas station, Royal Oak, Michigan. From Chester Liebs, *Main Street to Miracle Mile*, Boston: Little, Brown, 1985. Reprinted by permission.

have a tremendous impact on the way we analyze popular architecture. The development of designs and the evolution of styles in relationship to popular media beginning in the early nineteenth century remains largely unexplored. Ground-breaking research will attempt to place the client–architect–builder of popular vernacular architecture into a relationship with the system of mass media and communication.

Popular architecture and living people. Unlike the writing of architectural history based on limited documentation and vanished people, vernacular/traditional building scholars studying popular architecture will increasingly be confronted with more information and living people who have owned, built, designed, or lived in the architecture (see Figure 15). These new sources of information will necessitate new methods and adjustments, especially to historical vernacular scholarship. It is here that the behavioral sciences and contemporary cultural studies will have the greatest impact.

Industrialization and urbanization. When American vernacular scholarship opened its doors to the late nineteenth and twentieth centuries, in

FIGURE 15. Outdoor shrine maker, Carroll Gardens, Brooklyn, New York. From Thomas Carter & Bernard L. Herman, (Eds.), *Perspectives in Vernacular Architecture* (Vol. 3), Columbia, MO: University of Missouri Press, 1989. Reprinted by permission.

rushed the industrial revolution and urban America (see Figure 16). As twin topics, they have been uniformly avoided by vernacular/traditional building scholars who have concentrated on rural and domestic environments. Even when these topics have been addressed substantively, they are seldom portrayed as positive factors in the development of rural, preindustrial vernacular buildings and their environments (Stilgoe, 1982). This vernacular romanticism will have to be replaced by a new positivistic attitude toward industry and urbanism. Watch for the biggest gains to be made by researchers who can successfully interpret the relationship between industrial capitalism, consumerism, and the development of the American middle class.

Suburbanization. The vast phenomena of a suburbia is a theme that characterizes American popular building and environment values perhaps more than any other (Stilgoe, 1988; Figure 17). It is a topic in need of the finest vernacular/traditional architectural interpretation. The scholars who can harness this great summation of American culture will direct a major current of American architectural studies in the future.

FIGURE 16. Warehouse, Buffalo. From Reyner Banham, *A Concrete Atlantis,* Cambridge: MIT Press, 1986. Reprinted by permission.

FIGURE 17. Suburban houses (1947), Levittown, New York. From Diane Maddex (Ed.), *Built in the USA*, Washington, DC: Preservation Press, 1985. Reprinted by permission.

CONCLUSION

Whatever the various future directions for vernacular/traditional architecture studies, careful analytical building documentation will continue to form the bedrock for future research. The finest works of vernacular/traditional architectural research have been grounded in intimate, firsthand knowledge of the vernacular building (poor studies are invariably based upon an inadequate grasp of the building evidence and its environmental and cultural context). The significance of this relationship between vernacular research and building artifact goes to the very core of vernacular studies and offers a model for predicting future advances. American vernacular architecture and its study has often been characterized by a profound material objectivity—a love for building as building. It is this spirit, an American vernacular spirit, that seems to inform vernacular scholarship and will probably continue to do so.

REFERENCES

Allsopp, B. (1970). *The study of architectural history.* New York: Praeger.
Banham, R. (1986). *A concrete Atlantis:* Cambridge, MA: MIT Press.

Bloom, A. (1987). *The closing of the American mind*. New York: Simon & Schuster.

Bronner, S. J. (1983). "Visible proofs": Material culture study in American folkloristics. *American Quarterly, 35*(3), 316–338.

Bronner, S. J. (1985). American material culture and folklife: A prologue and dialogue. In S. J. Bronner (Ed.), *American material culture and folklife* (pp. 1–20). Ann Arbor, MI: University of Michigan Research Press.

Ching, F. D. K. (1979). *Architecture: Form, space & order*. New York: Van Nostrand Reinhold.

Cooper, C. C. (1975). *Easter Hill village: Some implications for design*. New York: Macmillan.

Cummings, A. L. (1979). *The timber framed houses of Massachusetts Bay, 1625–1725*. Cambridge, MA: Harvard University Press.

Fletcher, B. (1956). *A history of architecture on the comparative method*. New York: Scribner.

Foucault, M. (1980). *Power/knowledge*. New York: Pantheon.

Gans, H. J. (1974). *Popular culture and high culture*. New York: Basic Books.

Geertz, C. (1973). *The interpretation of cultures*. New York: Basic Books.

Glassie, H. (1968). *Patterns in the material folk culture of the Eastern United States*. Philadelphia: University of Pennsylvania Press.

Glassie, H. (1974). The variation of concepts within tradition: Barn building in Otsego County, New York. In H. J. Walker & W. G. Haag (Eds.), *Geoscience and Man* (Vol. 5, pp. 205–232). Baton Rouge: Louisiana State University.

Glassie, H. (1975). *Folk housing in Middle Virginia: A structural analysis of historic artifacts*. Knoxville: University of Tennessee Press.

Glassie, H. (1977). Meaningful things and appropriate myths: The artifact's place in American studies. In J. Salzman (Ed.), *Prospects: An annual of American cultural studies* (pp. 1–55). New York: Burt Franklin.

Glassie, H. (1982). *Passing the time in Ballymenone: Culture and history of an Ulster community*. Philadelphia: University of Pennsylvania Press.

Goat, L. G. (1989). Housing the horseless carriage: America's early private garages. In T. Carter & B. L. Herman (Eds.), *Perspectives in vernacular architecture* (Vol. 3, pp. 62–73). Columbia: University of Missouri Press.

Gombrick, E. H. (1979). *The sense of order*. Ithaca, NY: Cornell University Press.

Hart, J. F. (1975). *The look of the land*. Englewood Cliffs, NJ: Prentice-Hall.

Himmelfarb, G. (1959). *Darwin and the Darwinian revolution*. Garden City, NY: Doubleday.

Hirshorn, P., & Izenour, S. (1979). *White towers*. Cambridge, MA: MIT Press.

Hubka, T. C. (1984). *Big house, little house, back house, barn: The connected farm buildings of New England*. Hanover, NH: University Press of New England.

Hubka, T. C. (1985a). In the vernacular: Classifying American folk and popular architecture. *The Forum: Bulletin of the Society of Architectural Historians, 7*, 1–2.

Hubka, T. C. (1985b). Just folks designing: Vernacular designers and the generation of form. In D. Upton & J. M. Vlach (Eds.) *Common places: Readings in American vernacular architecture* (pp. 426–432). Athens: The University of Georgia Press.

Ivy, R. A., Jr. (1989, August). As it stands, a school balance. *Architecture*, pp. 42–49.

Jencks, C., & Baird, G. (1970). *Meaning in architecture*. New York: George Braziller.

Jordan, T. G. (1985). *American log buildings*. Chapel Hill: University of North Carolina Press.

Kleinsasser, W. (1979). *Some essential concerns: Experiential considerations in environmental design*. Eugene: University of Oregon, Department of Architecture.

Kostof, S. (1985). *A history of architecture: Settings and rituals*. New York: Oxford University Press.

Kowinski, W. S. (1985). *The malling of America: An insider look at the great consumer paradise*. New York: William Morrow.

Langer, S. K. (1953). *Feeling and form: A theory of art*. New York: Scribner.

Lawrence, R. J. (1983). The interpretation of vernacular architecture. *Vernacular Architecture, 14,* 19–28.

Liebs, C. H. (1985). *Main Street to miracle mile: American roadside architecture.* Boston: Little, Brown.

Lounsbury, C. (1989). The structures of justice: The courthouses of Colonial Virginia. In T. Carter & B. L. Herman (Eds.), *Perspectives in vernacular architecture* (Vol. 3, pp. 214–226). Columbia: University of Missouri Press.

Marshall, H. W. (1981). *American folk architecture: A selected bibliography.* (Publications of the American Folklife Center, No. 8). Washington, DC: American Folklife Center, Library of Congress.

Nelson, D. E. (1982). "Right nice little house[s]": Impermanent camp architecture of the American Civil War. In C. Wells (Ed.), *Perspectives in vernacular architecture* (Vol. I, pp. 131–142). Columbia: University of Missouri Press.

Noble, A. G. (1984). *Wood, brick, and stone* (Vol. 2). Amherst: University of Massachusetts Press.

Pearson, M. (1989). Annual meeting–Boston, 1990. *Newsletter: The Society of Architectural Historians, 33*(2), 1–16.

Pevsner, N. (1976). *A history of building types.* Princeton: Princeton University Press.

Rapoport, A. (1983a). Development, change and supportive design. *Habitat International, 7,* 249–268.

Rapoport, A. (1983b). Environmental quality, metropolitan areas and traditional settlements. *Habitat International, 7*(314), 37–63.

Riley, R. B. (1987). Vernacular landscapes. In E. H. Zube & G. T. Moore (Eds.), *Advances in environment, behavior, and design* (Vol. 1, pp. 213–225). New York: Plenum.

Roth, L. M. (1979). *A concise history of American architecture.* New York: Harper & Row.

Rudofsky, B. (1977). *The prodigious builders.* New York: Harcourt, Brace, Jovanovich.

Ruskin, J. (1884). *The seven lamps of architecture.* New York: Wiley.

Rykwert, J. (1981). *On Adam's house in paradise.* Cambridge, MA: MIT Press.

Sadler, J. T., Jr., & Sadler, J. D. J. (1981). *American stables: An architectural tour.* Boston: New York Graphic Society.

Saile, D. G. (1985). Many dwellings: Views of a pueblo world. In D. Seamon & R. Magerauer (Eds.), *Dwellings, place, and environment* (pp. 57–89). Dordecht, the Netherlands: Martinus Nijhoff.

St. George, R. B. (1986). "Set thine house in order": The domestication of the yeomanry in seventeenth-century New England. In D. Upton & J. M. Vlach (Eds.), *Common places: Readings in American vernacular architecture* (pp. 336–366). Athens: University of Georgia Press.

Sciorpa, J. (1989). Yard shrines and sidewalk alters of New York's Italian-Americans. In T. Carter and B. L. Herman (Eds.), *Perspectives in vernacular architecture* (Vol. 3, pp. 185–198). Columbia: University of Missouri Press.

Scruton, R. (1979). *The aesthetics of architecture.* Princeton, NJ: Princeton University Press.

Sloane, E. (1967). *An age of barns.* New York: Ballantine Books.

Stilgoe, J. R. (1982). *Common landscape of America, 1580 to 1845.* New Haven, CT: Yale University Press.

Stilgoe, J. R. (1988). *Borderland: Origins of the American suburb, 1820–1939.* New Haven, CT: Yale University Press.

Trachtenberg, M. (1988–1989). Some observations on recent architectural history. *Art Bulletin, 70*(2), 208–241.

Turner, P. V. (1984). *Campus: An American planning institution.* Cambridge, MA: MIT Press.

Upton, D. (1983). The power of things: Recent studies in American vernacular architecture. *American Quarterly, 35*(3), 262–279.

Upton, D. (1986). *Holy things and profane: Anglican parish churches in Colonial Virginia.* Cambridge, MA: MIT Press.

Upton, D. (1988, May). *American vernacular architecture.* Keynote address given at the annual meeting of the Vernacular Architecture Forum, Stanton, VA.

Upton, D., & Vlach, J. M. (1986). Introduction. *Common places: Readings in American vernacular architecture.* Athens: University of Georgia Press.

Ustunkok, O. (1988). Assessment of the vernacular: The role of physical, typal, morphological aspects. *Design Methods and Theories Journal, 22*(3), 844–864.

Venturi, R., Brown, D. S., & Izenour, S. (1977). *Learning from Las Vegas.* Cambridge, MA: MIT Press.

Vitruvius, P. (1960). *The ten books of architecture.* New York: Dover. (Original work published 1899)

Watkin, D. (1977). *Morality and architecture.* Oxford, England: Charendon.

Wells, C. (1986). Old claims and new demands: Vernacular architecture studies today. In C. Wells (Ed.), *Perspectives in vernacular architecture* (Vol. 2, pp. 1–11). Columbia: University of Missouri Press.

Wyatt, B. (1986). The challenge of addressing vernacular architecture in a state historic preservation program. In C. Wells (Ed), *Perspectives in vernacular architecture* (Vol. 2, pp. 37–43). Columbia: University of Missouri Press.

Zelinsky, W. (1973). *The cultural geography of the Eastern United States.* Englewood Cliffs, NJ: Prentice-Hall.

III

ADVANCES IN USER
GROUP RESEARCH

6

Homes for Children in a Changing Society

LOUISE CHAWLA

This chapter reviews research and design advances in housing for children and adolescents in developed nations. The places it focuses upon are the dwelling and the housing site—the places which form the primary matrix of their lives. Even when parents of infants and preschoolers seek child care, prevalent arrangements are in-home care and family day care. Across the span of a year, school-age children and adolescents spend more than four-fifths of their time out of school, which means primarily in or near the home. At its best, housing lends itself to the creation of settings where families can thrive. This chapter reviews recent attempts to define housing at its best.

In the late nineteenth century and early twentieth century, the quality of children's homes and neighborhoods was an important issue for progressive reformers; but as child study moved into the university laboratory, real-world settings became neglected. During the environmental movement of the 1970s, the quality of homes and neighborhoods received renewed attention. Since 1975, there has been a steady flow of relevant publications, including a number of reviews and edited collections upon which this chapter has built. This survey of these resources will emphasize work published in the last decade, with three primary purposes: to observe where intensive research has already been done; to identify issues that have been neglected; and to highlight design implications.

Louise Chawla • Whitney Young College, Kentucky State University, Frankfort, Kentucky 40601.

An important consideration in evaluating this literature has been the degree to which it has acknowledged the impact of contemporary social and economic changes upon children's experience. In industrialized countries, birth rates are declining and families are restructuring as a consequence of divorce or nonmarriage, resulting in fewer households with children and fewer people per household (Wattenberg, 1987). In the area of economics, rising land and building costs, decreasing government subsidization of housing, the entry of women into the paid labor force in increasing numbers, and high rates of poverty among families with children (Huttman, 1985) have combined to make the fit between contemporary environments and children's needs problematic. In keeping with these changes, this chapter will give attention to design, planning, and policy recommendations for housing for a variety of family forms.

METHODOLOGICAL ISSUES IN CHILD–ENVIRONMENT RESEARCH

A review of specific studies needs to be prefaced by a few general comments regarding the status of child–environment theory and methods. In 1980, the late Joachim Wohlwill noted promising signs of convergence between developmental and environmental psychology. The fields had come to the common conclusion that as people act upon their settings, their settings act upon them, shaping development. Wohlwill expected a partial confluence of these two disciplines to benefit research in two ways: developmentalists could contribute an effective integration of empirical facts with theory, and environmentalists a readiness to apply naturalistic methods in real-world settings. Wohlwill noted that an emerging ecological paradigm in developmental theory, such as that promoted by Bronfenbrenner (1979; Table 1), had created a framework within which to examine the nested socio*physical* systems within which a child's life is embedded; but he cautioned that in practice developmentalists had continued to stress social relationships at the expense of the physical environment. For research to be truly ecological, Wohlwill urged, it would need to incorporate more quasi-experimental designs in natural settings, to relate behavior to children's experiential histories, and to sample across a broad range of well-defined settings and conditions.

In child–environment research, opportunities to build and refine theory by comparing findings across settings and conditions have been hampered in the past by frequent weaknesses in research design and by uncoordinated strategies. G. Moore (1982) has faulted the field for its

Table 1. Guiding Principles of Ecological Research

1. The purpose of research is to generate findings relevant to social policy, which requires an understanding of natural situations.
2. Development is a transactional affair in which person and environment reciprocally act upon one another.
3. The environment can be conceived as a set of nested sociophysical systems that are themselves reciprocally related. Successive systems that shape a child's life are the immediate people, places, and things that the child encounters; informal and institutional networks that link settings, such as friendship networks that extend between home and school; settings which indirectly affect life, such as the parent's workplace or the local planning board; and the embracing culture of assumptions, policies, and institutional patterns.
4. The larger context mediates person–environment interactions, creating second-order effects. For example, the freedom to play in the home that a mother allows a child may be mediated by her own relationship with the father.

Note. From Bronfenbrenner, 1979.

infrequent use of well-controlled quasi-experimental methods. Weinstein and David (1987) and Van Vliet (1985) have described the field as fragmented, without consistent definitions or measures. In the words of Van Vliet (p. 64), "a novel measure is the rule, replication the exception," with the result that "there are few instances of validation or refutation of reported findings."

One solution to the problem of inconsistent measures is for a single researcher or research group to sample a range of settings in one study. Model work of this kind has been coordinated by Lynch (1977) in four international cities, by R. Moore (1986) in three British cities, by Berg and Medrich (1980) and Homel and Burns (1985) in a variety of neighborhoods in and around one city, and by Bjorklid (1985) and Mackintosh (1985) on diverse housing estates. Such work, however, requires high levels of dedication or funding. More replication of methods and measures across studies is needed.

The investigation of children's housing experience has been further complicated by the nature of its subject and the nature of its population. Family rights to privacy and noninterference in the home limit researchers' opportunities to control settings or make direct observations. Yet a recent review of child–environment methods by Ziegler and Andrews (1987) has underscored the importance of objective data, noting that parents' subjective ratings of the home have not always corresponded to objective measures. There is a tradition of home-based research with infants and preschoolers that has incorporated home inventories (Caldwell, 1968) and observations (White, Kaban, & Attanucci,

1979). Noteworthy strategies with older children have been introduced by Sebba and Churchman (1983), who related interview responses to apartment maps, and by Schiavo (1988), who asked adolescents to discuss home photographs that they had taken themselves.

Another complication is that methods must be age appropriate. This issue has been well covered by Ziegler and Andrews (1987), who have surveyed research methods used in the past, distinguishing those appropriate to different stages of development. Noting the strengths and weaknesses of each method, they have recommended multimethod approaches. They have also identified the critical issue that faces anyone who studies children.

> Since investigators are virtually always adults, their basic understanding of objects and events is likely to be qualitatively similar to that of adult subjects, but qualitatively different from that of child subjects. In this sense, an adult investigator of children's environmental interactions is akin to an ethnologist working in a culture foreign to his or her own, and therefore faced with having to build in safeguards against culturally biased interpretation of the data. (p. 302)

In view of young children's inarticulateness, Ziegler and Andrews have observed that quantitative methods, such as behavior checklists, must remain an important component of research. The preceding caveat, however, requires that they be balanced by qualitative insights, that empirical data be matched by phenomenological description that clarifies the meaning of the data from both the researcher's and the child's perspective. Multimethod studies of this kind have been pioneered by Hart (1979) and R. Moore (1986), who have combined objective measures of the environment with child maps, models, and interviews.

Because the ultimate goal of design research is the preservation or creation of optimal places for users, it is important to relate children's experience to the attitudes and behavior of their parents and the neighbors, landlords, building managers, design professionals, developers, and planning board officials who shape their spaces. Action-oriented research that attempts to incorporate children into environmental decision making will be reviewed in a later section on children's participation in design.

THE HOME INTERIOR

There are three main bodies of research on children's housing experience. First, and most copious, is research on the effects of stimulus levels in the home: noise, density, and opportunities for sensorimotor stimulation. This literature has paid particular attention to the effects of stimulus levels upon cognitive development, with lesser attention to their effects upon motor and social development, and it has dealt primarily with infants and young children, secondarily with school-age children, and almost not at all with adolescents. Second in volume are housing evaluations and recommendations that have examined the suitability of different dwelling layouts for families with children. Smallest in number and most fragmented in method are a few studies that have begun to explore the qualitative meaning of home for children.

Stimulus Levels in the Home

The large literature on stimulus levels in the home will not be reviewed here because thorough cumulative reviews already exist by Parke (1978), Wachs and Gruen (1982), and Wohlwill and Heft (1987). Papers on the impact of density on children have been collected by Wohlwill and Van Vliet (1985). Dejoy (1983) has analyzed the findings on children and noise. The conclusions of these reviews will be briefly summarized, with an emphasis upon their implications for design.

The following major relationships have been found. With regard to opportunities for sensorimotor stimulation, the variety, complexity, and responsivity of the objects made available to young children in the home have correlated with multiple measures of cognitive and motivational development (Parke, 1978, pp. 45–50; Wachs & Gruen, 1982, pp. 42–52). Some of these early opportunities have shown enduring effects upon I.Q. scores (Wachs & Gruen, 1982). Floor freedom and an absence of physical barriers and parental prohibition against infants' exploration have also been positively related to cognitive development (Wohlwill & Heft, 1987, pp. 296–297).

These findings are not surprising, considering the importance of free movement (Held & Hein, 1963) and object manipulation (Piaget, 1952) for sensory coordination and cognition. The importance of opportunities for free exploration is attested to by a major longitudinal study of early development by White et al. (1979), in which home observations revealed that one- to three-year-olds spend approximately 80–90 percent of their waking hours interacting with their physical, rather than social, environment. Caretakers who fostered competence, the study showed, structured the environment for safety and maximum access, and served as consultants to help the child do things independently. Because children of this age were observed to spend more waking hours in the kitchen than in any other room, researchers stressed the importance of making floor-level kitchen cabinets available for exploration and making this room a safe play space.

In contrast to sensorimotor stimulation, noise has an adverse effect on development. Noise in the home has been found to be inversely related to measures of early language development in infants; and both interior and exterior noise has been shown to impair information processing and language achievement in older children (DeJoy, 1983; Parke, 1978, pp. 61–64; Wohlwill & Heft, 1987, pp. 293–295). Residential density, which may be taken as an indirect measure of crowding, noise, and activity levels in the home, has shown similar impacts. High person-per-room densities are negatively associated with infant cognitive development; among older children, they are negatively related to information-processing skills and vocabulary development and positively related to aggression, anger, and acting out (Wohlwill & Heft, 1987, p. 295). Anger and aggression under these conditions may reflect parents' responses to crowding, which appear to be more punitive child-rearing and looser supervision of children's whereabouts outdoors (Aiello, Thompson, & Baum, 1985, pp. 110–111).

The preceding findings have some clear design implications. To protect children from noise that is not under their control (and adults from noise that is), adequate soundproofing is a priority between rooms, between dwelling units, and between the dwelling and the outdoors in noisy neighborhoods. To enable infants and preschoolers to play with a variety of responsive objects within the sight and hearing of parents, but not underfoot, another priority is space in view of parents' work areas that can be easily managed for play. As a general rule, children need to be able to negotiate optimal levels of stimulation for themselves by alternately seeking proximity or withdrawal from people and things. For infants and preschoolers, Wachs (1979) has argued that a "stimulus shelter"—a room or corner where they can withdraw from household activity when they desire—is an important correlate of early cognitive development. A similar need among children is suggested by the finding of Michelson (1968; cited in Parke, 1978) that third graders who could study in a separate room free from distractions had higher spelling, language achievement, and creativity scores than those without such a retreat, after controlling for parents' education, occupation, and income. Research by Parke and Sawin (1979) and Wolfe (1978) has shown that, by early adolescence, children value opportunities to appropriate a room and to control intrusions as essential aspects of privacy.

Housing Design

In addition to the preceding general recommendations, more specific recommendations have been derived from postoccupancy evaluations of low- and middle-income housing. Zeisel and Welch (1981) and

Cooper-Marcus and Hogue (1977) have summarized more than a decade of evaluation results. Pollowy (1977) has integrated evaluation results with a review of developmental norms from infancy through middle childhood. Zinn (1980) has considered family needs in the context of a social history of the home. Despite the unquestionable value of this work in synthesizing user needs research, it must be kept in mind that recommendations as to how housing can best serve both parents and children remain educated inductions from limited data. Evaluations have been primarily based upon interview surveys with parents, reflecting parents' concerns for order and child supervision, and the observation of children's outdoor play behavior. There has not yet been research which has simultaneously surveyed parents' and children's responses to different dwelling designs, or which has compared interview reports with the actual observation of families' interior space use.

The preceding studies have derived similar design recommendations. Table 2 presents major recommendations that recur in more than one of these studies. The general ideal can be seen to be an interior adaptable to the changing stages of the family life cycle, that allows parents and children to pursue common, parallel, or private activities with minimal conflict.

Figure 1 shows model plans which have incorporated some of these recommendations in a government-subsidized West German housing project. A large kitchen/family room and flexible bedrooms have been achieved without increasing the standard floor area or cost. Figure 2

TABLE 2. Design Recommendations for Housing Interiors for Families with Children

1. Create a kitchen/family room where young children can play in proximity to parents.
2. Provide a safely enclosed semipublic or private balcony or yard for young children's play.
3. Plan for flexible bedrooms that can be partitioned to offer different degrees of shared space or privacy among siblings.
4. Whenever possible, provide a spare room that can be adapted for play, hobbies, home-based work, study, or guests as family needs change.
5. If two full bathrooms are not feasible, two partial bathrooms are preferable to one large family bathroom.
6. Ensure soundproofing between dwelling units, between rooms, and between the interior and exterior in noisy neighborhoods, through insulation, vibration breaks between walls and floors, and the use of closets and hallways as buffers.
7. Provide a mud room at the most used entrance.
8. Provide adequate storage inside the dwelling and in a locked outdoor shed or communal storeroom.

Note. From Cooper-Marcus & Hogue, 1977; Pollowy, 1977; Zeisel & Welch, 1981; Zinn, 1980.

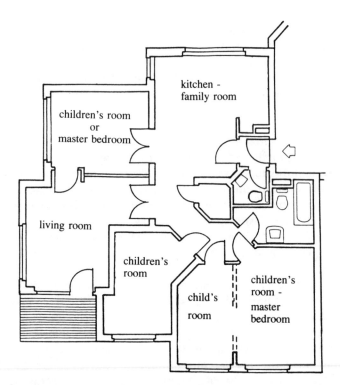

FIGURE 1. Model floor plans for government-subsidized housing in Frankfurt am Main, West Germany. In a pilot study, residents with two to four children especially liked the flexible rooms, the large kitchen/family area, and the greater play space. These innovations were achieved within existing standard floor areas (design by I. Rojan-Sandvoss). From A. Flade, "Evaluation of Housing Floor Plans with Regard to Meeting Family Needs," *Children's Environments Quarterly,* 1986, 3. Reprinted by permission of the author.

shows a plan derived from a participatory design project involving low-income, single-parent women in Providence, Rhode Island. Given budget constraints, they too have chosen to maximize a kitchen/family room at the expense of space in the living room and bedroom.

The most frequently stressed recommendation in the above studies is adequate floor space for play. Contrary to this need, when Gaunt (1980) interviewed a sample of 120 families with children between the ages of two and seven regarding indoor play patterns, she found that children spent much more time in quiet, passive play than in active, creative play. Although families in larger living quarters reported more active play, this general pattern remained constant.

Whether or not children engage in creative indoor play ultimately

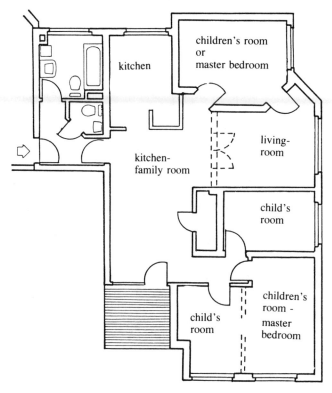

FIGURE 1. (*Cont.*)

depends upon whether or not parents encourage it; but the environment sends implicit messages to both parents and children regarding its appropriateness. Families must commonly cope with multiple barriers to free play which suggest that it is *not* expected: no open floor space for this purpose, no storage space for toys and supplies, poor soundproofing, difficult-to-clean surfaces. These stresses were evident in a study by Rubenstein and Howes (1979), who observed matched samples of 18-month-olds in homes and in a day-care center. Finding that infants at home received more reprimands and cried more than their day-care peers, they concluded that the typical dwelling is ill-suited for child-rearing. The primary sources of friction in the home that they identified were that parents found themselves isolated, without other caregivers' support, and that rooms were not child oriented.

This finding suggests that child care center design recommendations should be studied for their applicability to the private dwelling. An

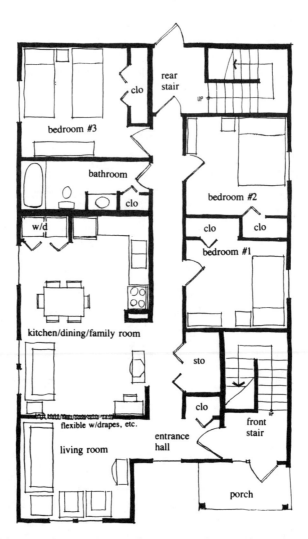

FIGURE 2. Women's Development Corporation in Providence, Rhode Island, and 25 low-income, single-woman heads of households spent two years planning the development of 100 units of housing. After months of participatory design exercises, they created a prototypical design program for an apartment responsive to urban site constraints, space and cost limitations, and a need for three bedrooms. From this program, Nancy Santagata designed the above plan. It puts maximum square footage in a dining/kitchen/family area and sets aside the living room as a small passive activity space. A postoccupancy evaluation by Myrna Breitbart has supported this design concept. Information on the design process and evaluation can be obtained from Women's Development Corporation, 861A Broad Street, Providence, Rhode Island. Picture courtesy Women's Development Corporation.

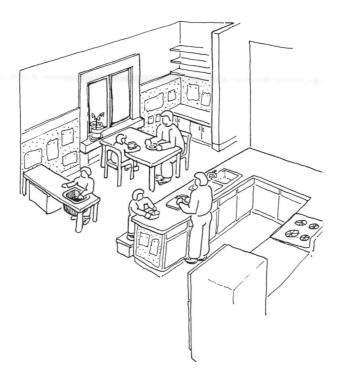

FIGURE 3. Adaptations to the kitchen to support children's playing and learning (drawing and design by Joel Shack). From L. C. Johnson, J. Shack, & K. Oster, *Out of the Cellar and into the Parlour*, Ottawa: Canada Mortgage & Housing Corporation, 1980. Reprinted by permission of Canada Mortgage & Housing Corporation and the artist.

advantage of this approach is that it taps a rich empirical literature. Of particular relevance is the work by Moore, Lane, Hill, Cohen, and McGinty (1979, #921) and Johnson, Shack, and Oster (1980) regarding the adaptation of the home for family day care, and work by Greenman (1987) and Olds (1987) regarding the creation of homelike institutions. All of the above authors have made detailed practical suggestions for the appropriate scaling of the environment for children and for the creation of easily maintained activity centers. Figure 3 shows how children can be accommodated in the kitchen—the room that toddlers were observed to use most intensively in the study by White *et al.* (1979).

Model work by architects and interior designers reproduced in McGrath and McGrath (1978) and Bevington (1987a) has demonstrated the particular usefulness of alcoves, lofts, and room dividers to create spaces for children. Most of this work has been commissioned by well-

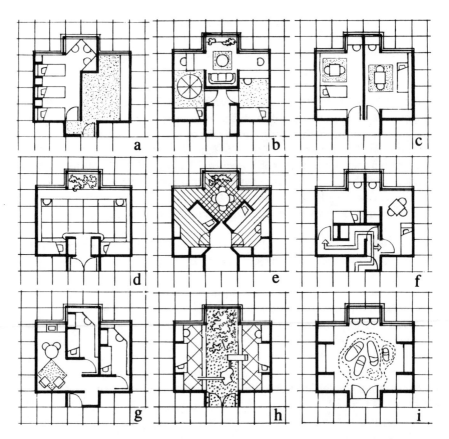

FIGURE 4. A bedroom with alcoves that supports flexible layouts according to a "quilt-plan." Out of 35 suggested variations, the plans reproduced here offer different degrees of togetherness or aloneness (a–c), simplicity or complexity (d–f), and fixity or flow (g–h). From C. B. Bevington, "A Quilt-Plan," *Interior Design*, May 1987a. Reprinted by permission of the author.

to-do families; yet its space-stretching techniques are applicable to low-cost housing as well. Figure 4 illustrates the flexibility of a bedroom with alcoves in contrast to a rectangular room.

Although the home may be made more child oriented, this improvement does not address the other problem identified by Rubenstein and Howes: that parents often find themselves isolated among their children in the private home. For children in small families, the result is also isolation from their peers. Histories of housing have traced the roots of this isolation to nineteenth-century industrialization and an ideology of "separate spheres," which assigned women and children to the domestic realm at a safe distance from the competitive wage-earning world of men (Franck, 1985; Hayden, 1984). Franck (1985) has noted that as separation was enforced through segregated land-use zoning and detached private housing, the burden of housework fell on women. At the same time, children and teenagers were left few opportunities to contribute to household maintenance or to observe or participate in adult roles outside the home, and made dependent upon parents for transportation to friends and community resources.

As more and more women have become wage earners and heads of households, environments that embody separate spheres have come under increasing criticism (Peterson, 1987). As mismatches between family needs and existing housing are identified, the great current challenge has become to design housing that will reflect actual family diversity, and to effect the zoning changes and economic strategies that will make it possible to construct it. Hayden (1982, 1984) has reviewed the checkered 100-year history of comprehensive schemes to incorporate shared dining, housekeeping, nursing care, child care, and play facilities within multifamily housing. Frank and Ahrentzen (1989) have collected reports of recent projects. A first step toward providing needed services is to supplement the private dwelling with communal indoor play rooms, supervised playgrounds, drop-in centers, and child care centers. Figures 5 through 7 reproduce plans that incorporate some of these services in housing for homeless families and unwed teenage mothers, freeing parents to pursue their education and employment.

In building housing, location and management are as important as design. Cooper (1975) and Sarkissian and Doherty (1987) have documented the physical deterioration of the housing site and the social stresses that result when female-headed households are concentrated in large new developments. Here abnormally high densities of children and adolescents are cut off from community services, youth employment opportunities, and balanced contacts with older generations.

The importance of generational balance is underscored by a carefully controlled study of vandalism by Wilson (1980; cited in Harvey, 1982), in which child density turned out to be the single most important factor in explaining variations in vandalism on housing estates. In a review of preventive measures against vandalism, Cooper Marcus and Sarkissian (1986, pp. 280–285) have cautioned that whenever densities exceed 30 children per acre, special recreation provisions are necessary. In view of these problems, Figures 5 through 7 illustrate schemes to integrate a small number of families into established neighborhoods with basic services, and to provide resident management.

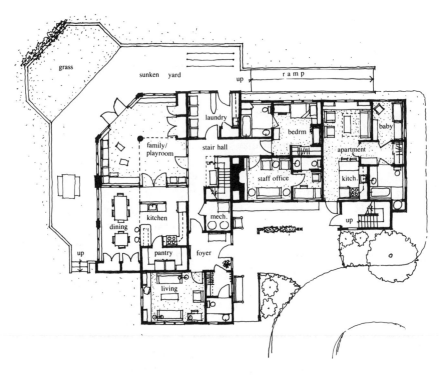

Figure 5. Plan for a Yarmouth, Massachusetts, teen mother's residence. The first floor of a three-story house for eight residents and staff. Entering residents contract to complete high school or vocational training after the birth of their child. During their first year, mothers occupy bedrooms and share cooking and dining facilities, as they learn nutrition, health care, and child care. In their second year, they graduate to private efficiency apartments. Programming by Welch & Epp Associates, Boston. Picture courtesy Prellwitz/Chilinski, Architects, Cambridge.

The Meaning of Home

In addition to physical shelter, Hayward (1975) has proposed that a home serves as a place for privacy and refuge, for social affiliation, for personalization and self-identity, for activity, and for continuity. How housing serves these different dimensions of a home, from children's own perspectives, has barely begun to be explored. Initial insights have been gathered through interviews (Csikszentmihalyi & Rochberg-Halton, 1981; Ladd, 1972; Sweaney, Inman, Wallinga, & Dias, 1986), drawings (Filipovitch, Juliar, & Ross, 1981; Neperud, 1975) and essays

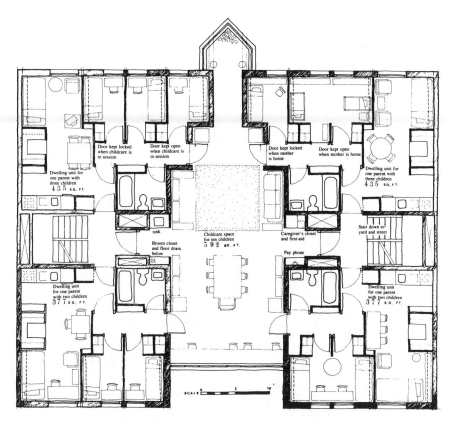

FIGURE 6. Proposed housing for the homeless mother and child. The plan is conceived to occupy the second and third floors of a four-story New York apartment building, thus mixing subsidized and unsubsidized families. Each floor accommodates four families. Parents' quarters may be locked for privacy. Children's private quarters open into a central child care area to eliminate the need for building and maintaining public toilets, cubbies, and nap areas. This basic plan can be elaborated to serve the needs of dual-career families as well. From C. B. Bevington, "Housing the Homeless Mother and Child," *Women & Environments*, Fall, 1987b. Reprinted by permission of the author.

(Neperud, 1975). The most reliable work has combined interviews or drawings with objective measures in the form of photographs (Schiavo, 1988) or floor plans (Sebba & Churchman, 1983).

Studies that have asked children to draw or describe their home environments suggest that important places vary with age, gender, and housing type; but because no one has controlled for all three factors, or replicated measures, relationships remain unclear. It appears, however,

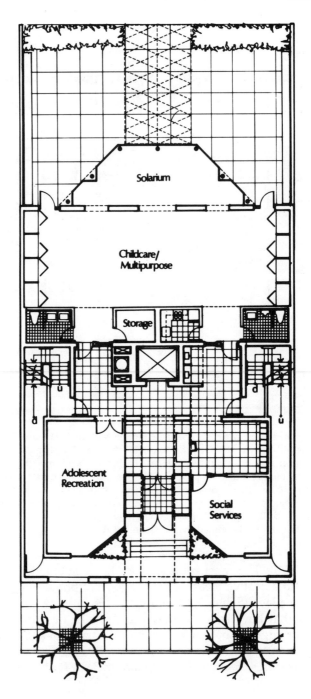

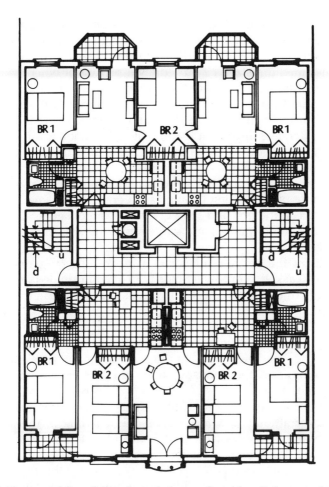

FIGURE 7. The ground floor (left) and a typical upper floor (above) of proposed transitional housing in New York that has been designed to serve as a neighborhood asset. To facilitate the social integration of residents into the community, the ground floor plan offers much needed child care, adolescent recreation, and social services to neighbors as well as residents. The back of the building presents a choice of sheltered, semisheltered, or open play spaces. A corner with sand, bark, or pea gravel has yet to be added in the back to provide manipulable ground material and a safe foundation for play equipment. Above, a plan of the second through fifth floors flexibly accommodates families of different sizes through self-contained one or two bedroom apartments with "swing" bedrooms, and through four bedroom apartments consisting of two family suites joined by a shared living/dining space. The sixth floor (not shown) provides a laundry room with an adjacent roof terrace; the cellar contains family storage lockers. Design by Conrad Levenson & Marvin Meltzer. Picture courtesy of Levenson Meltzer Neuringer, Architects and Planners, New York.

that children invest greatest attention in the home if they are very young or adolescent, female, or live in a single-family suburban house.

When Filipovitch *et al.* (1981) asked fourth, seventh, and twelfth graders in a medium-sized city to "draw a picture of where you live," responses shifted with age from the dwelling, to the community, to the dwelling again. Fourth graders drew a world that revolved around their house and yard. Seventh graders drew an explorer's world in which the house was often less important than the neighborhood at large. Twelfth graders returned to the house as their center of attention, with carefully individualized and personalized pictures.

Neperud (1975) found the same pattern when he analyzed drawings of favorite places by first through sixth graders in a small city and a suburb. Between the first and sixth grade, he found a general decline in the representation of houses and yards and a dramatic increase in outdoor scenes.

Neither of the above studies compared results by gender. When Schiavo (1987) asked 8- to 18-year-olds who lived in single-family suburban homes to draw, photograph, and discuss "any place in the home or neighborhood that is important to you," only boys showed a shift of attention from home, to neighborhood, to home again. Middle-school boys listed the greatest number of important neighborhood places, girls of the same age the fewest. As a group, suburban subjects showed an increasing emphasis upon housing interiors at the expense of the yard or neighborhood.

These differences point to distinct interests and opportunities that deserve closer attention. A focus upon the dwelling in the suburban study, for example, may reflect real housing advantages, a lack of neighborhood attractions, or both factors together. A survey of residential satisfaction by Michelson (1977, p. 285) suggests that this focus reflects housing form. In this survey, children in suburban single-family homes primarily listed dwelling characteristics as sources of satisfaction, in contrast to children in suburban apartment buildings, who primarily listed site characteristics and neighborhood facilities. Among girls, a greater emphasis upon the house conforms with a review of research on children's home range by Saegert & Hart (1978), which indicated that young children are restricted to the house and its immediate exterior; as they grow, boys are allowed a larger free range, whereas girls know the home and close-to-home spaces more intimately. A focus on interiors by both urban and suburban adolescents accords with the early suggestion by Ladd (1972) that adolescents as a rule center their attention upon interior space and personal objects.

Within the interior, consistent preferences have been found. According to qualitative research, beginning as early as age three, children highly value having a room of their own. In interviews, children who have had the advantage of their own room have named it as the primary place where they feel most at home or most like to be (Csikszentmihalyi & Rochberg-Halton, 1981; Ladd, 1972; Schiavo, 1988; Sweaney et al., 1986), which they can personalize and which represents them (Schiavo, 1988; Sebba & Churchman, 1983), where they are most likely to keep treasured possessions (Csikszentmihalyi & Rochberg-Halton, 1981), and where they retreat when they are upset or want to be undisturbed (Ladd, 1972; Sebba & Churchman, 1983; Sweaney et al., 1986). In interviews with 82 middle-class families in metropolitan Chicago (Csikszentmihalyi & Rochberg-Halton, 1981), children gave the greatest importance to their bedroom, parents the least, and grandparents an increased importance again.

The significance of the bedroom depends upon its size and privacy, however. Respondents with a shared bedroom have been less likely to name it as a favorite place (Schiavo, 1988) or as the place where they go to be alone (Ladd, 1972) or to read, play, or do homework (Sebba & Churchman, 1983). They have been much less likely to appropriate it as personal territory when they have shared a small room rather than a large room (Sebba & Churchman, 1983). These results are consistent with Swedish parents' reports summarized by Gaunt (1987), who found that children with separate bedrooms engaged in more varied, imaginative play and entertained more playmates than children with shared rooms. (See Figures 1 and 4 for bedroom designs that can flexibly accommodate both sharing and privacy among siblings.)

Other characteristics of children's home evaluations are more positive, emotional descriptions by girls than by boys (Csikszentmihalyi & Rochberg-Halton, 1981; Schiavo, 1988); a keen awareness of levels of maintenance by adolescents in run-down housing (Ladd, 1972); and an emphasis upon the instrumental value of places and things as opportunities to *do* something (Csikszentmihalyi & Rochberg-Halton, 1981; Neperud, 1975). The importance of opportunities for action is highlighted by two early studies that invited children and young adolescents to draw and model housing of their choice (Barbey, 1974; Thornberg, 1974). The results provided for adventurous locomotion via tunnels, ropes, ladders, and staircases; a tendency to merge rooms into circulation spaces; an inventive vocabulary of corners, curves, and variable ceiling heights; and prominent accessory spaces such as a music room, secret chambers, a pet farm, an orchard, and a subterranean lake. In the absence of real housing that can match these ideals, the nearest substitute is space where children can improvise these imaginary worlds for themselves.

ACCESS TO THE OUTDOORS: TRANSITIONAL SPACES

Whether the home forms a cage, a refuge, or a restorative base of operations for outward expansion depends not only upon its own qualities, but upon the qualities of the world beyond its walls. Between the private interior and the public world at large is a critical transitional margin: the access areas, yards, courtyards, and streets adjoining the home. In contrast to the relative scarcity of information regarding children's adjustments to housing interiors, an extensive tradition of evaluation research has documented their heavy use of this threshold between the public and private domain (see reviews by Cohen, Hill, Lane, McGinty, & Moore, 1979, #208, #603; Cooper-Marcus & Sarkissian, 1986, pp. 107–141; Pollowy, 1977, pp. 33–44; Zeisel & Welch, 1981, pp. 32–45, 79–80; and recent research by Churchman, 1980; Francis, 1985; Hart, 1979; R. Moore, 1986; Shack, 1987).

One significance of home-based territory that this literature has documented is that it is heavily used by children of all ages, whether they live in single-family or multifamily housing, in cities, suburbs, or rural areas. It offers secure access to the outdoors for young children who must remain under a caretaker's eye, and convenient access for older children who tend to fit snatches of outdoor play between homework, chores, mealtimes, and TV watching. It composes the greatest part of what Moore and Young (1978) have defined as the "habitual range" of children's daily outdoor activity.

Another significance of this close-to-home space is that here children can negotiate gradual independence. Its importance is illuminated by a model of outdoor access by Hart (1978). According to Hart, access to outdoor resources is "a dynamic adaptive process," "a three-way negotiative process between caretaker, child, and the environment" (p. 387). The three sides that families must balance are the child's concurrent needs to experiment, explore, and yet feel secure; caretakers' protectiveness; and actual attractions, risks, and dangers in the environment. The model distinguishes unpredictability and risk—which are inevitable if a child is going to enjoy adventure and test its growing competence— from danger, which defies learning given a child's age-related limitations. Noting that the dynamics of accessibility have been changing as parents have shifted more of their caretaking responsibilities to child

TABLE 3. Site Design Recommendations for Home-Based Play

1. Anticipate that children will play throughout the site, not only in playgrounds.
2. Plan for young children's play in spaces immediately adjoining the dwelling that are within sight and call from indoors.
3. Provide some hard surfaces for wheeled toys.
4. Provide a partially covered play area.
5. Partially screen an area for messy play.
6. Use grade changes, vegetation, and low walls to define activity areas.
7. Provide steps, low walls, or benches for adults' and adolescents' sitting.
8. Preserve distinctive natural site features such as trees, rocks, slopes, and water.
9. Set aside a wild area where children can explore nature and create their own private places.
10. Make loose materials available for creative play.
11. Orient play areas for shelter from wind, rain, and excessive sun.
12. Involve residents in the cooperative management and maintenance of communal spaces.

Note. From Canada Mortgage & Housing Corporation, 1978, 1979; Cohen, Hill, Lane, McGinty, & Moore, 1979; Cooper-Marcus & Sarkissian, 1986; Pollowy, 1977; Verwer, 1980; Zeisel & Welch, 1981.

care substitutes, as fears of traffic and crime have increased, and as higher densities have resulted in more restrictive control of the landscape by adults, Hart has described the challenge facing child and environmental professionals as "to consider how to guarantee children in the future accessibility to an environment without danger but one which provides unpredictability (or risk) which is the basis of adventure" (p. 387).

Drawing upon the results of nearly 100 postoccupancy evaluations, Cooper-Marcus and Sarkissian (1986) have recently compiled a comprehensive collection of site design guidelines for multifamily housing, with close attention to the needs of children and adolescents. Major recommendations made in this collection and other reviews are reprinted in Table 3. Rather than address details of site design that have already been well covered, this section will focus upon three contemporary challenges to children's access to outdoor adventure without danger: restrictive planning; high-rise housing; and children left in self-care.

ACCESS TO DIVERSE SPACES

At their best, recommendations for the design of children's close-to-home environments have integrated a review of developmental norms with a review of children's actual site use (Canada Mortgage & Housing Corporation, 1978, 1979; Cohen et al., 1979; Cooper-Marcus & Sarkissian, 1986; G. Moore, 1985; R. Moore, 1986; Pollowy, 1977; Verwer, 1980).

These reviews have come to two consistent conclusions: that the environment needs to support the whole child's physical, social, emotional, and cognitive development; and that the entire site needs to be designed for children's use. A corresponding consistent conclusion is that a playground, by itself, is never adequate. These reviews have noted that most playgrounds one-sidedly emphasize gross motor development, and that designated playgrounds only receive a small part of children's actual site use. Rather than playgrounds, research has shown children's heavy use of streets and sidewalks, followed by yards and undeveloped wild areas (G. Moore, 1985). These results are ironic, because official planning has traditionally concentrated upon the construction of playgrounds in an attempt to get children out of streets, neighbors' yards, and "waste" places. Empirical observations indicate that these are the places where children prefer to be.

The attempt to segregate family life from the world of paid work, whose impact on housing design has been reviewed, lies at the heart of current criticisms of site design as well. The advent of the automobile age in the 1920s made it possible to remove families from commercial and cultural centers, and all the more necessary to create protected spaces. Hayden (1984) has reviewed the history of residential planning in response to the automobile. The ideal of segregation and safety was promoted to be low-density suburbs where children could play on private lawns on quiet streets. On medium- and high-density estates, this ideal was approximated by clustering housing complexes around internal yards and pathways, excluding through traffic to the periphery.

For preschoolers and young school-age children, for whom a sandbox or a wading pool may represent high adventure, these solutions may work successfully. In multifamily housing, Barry (1982), Cooper-Marcus (1974), Mackintosh (1985), and Shack (1987) have recorded young children's heavy use of semipublic internal pathways, playgrounds, and courtyards, especially when they are observable from dwelling windows. McGinty, Cohen, and Moore (1982; see Figure 8) have shown how private yards can be conveniently adapted to preschool play. Yet there are several drawbacks to these solutions. Fewer and fewer mothers remain at home to watch their children from kitchen windows. As suburbs have become built up, many once quiet streets have come to support heavy traffic. For older children and adolescents, protected residential enclosures have little to offer (Keller, 1981; Popenoe, 1977; Sarkissian & Doherty, 1987). In the United States, there is evidence that low-density suburbs, which best fit traditional housing ideals, may be particularly restrictive, isolating environments for children (Berg & Medrich, 1980).

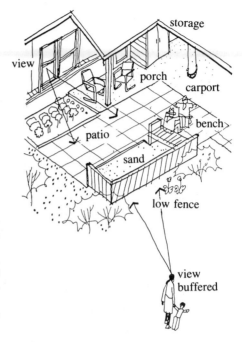

FIGURE 8. The private outdoor environment planned for home-based play. A small area can incorporate sheltered and open-air play, with grass, pavement, and sand surfaces in the view of caretakers either indoors or outdoors (drawing by Tim McGinty). From T. McGinty, U. Cohen, & G. T. Moore, *Play Environments*, final report to the U.S. Dept. of the Army, Office of the Chief of Engineers, Washington, DC (TM 5-803-11), 1982. Reprinted by permission of the artist.

The first step to the creation of more diverse environments must be the revision of restrictive zoning policies (Ritzdorf, 1986, 1987). Assuming zoning changes, Hayden (1984, pp. 186–191; reviewed in Peterson, 1987) has proposed that some suburban blocks be relandscaped to create central internal greens. These internal greens can make suburban living affordable and convenient for low-income, single-parent, and dual-career families by accommodating accessory apartments and services, including day care, along their margins; and they can provide a protected communal space for gardening, outdoor sitting, and play. By allowing accessory apartments, the plan may make it possible for a divorced mother to maintain the family home, or for grandparents to live in proximity to grandchildren. The plan also serves adolescents by creating close-to-home opportunities for after-school work.

Another approach to the creation of accessible, inviting outdoor space is to designate residential streets for pedestrians. Noting children's heavy use of streets for play, socializing, and watching the world go by, R. Moore (1987) has argued that residential streets should be conserved as *de facto* playgrounds. In the Netherlands, West Germany, and Scandinavia, this purpose has been achieved through the *woonerf*, a

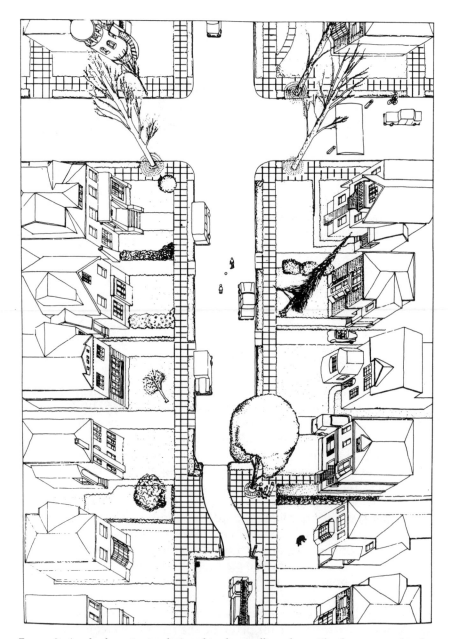

FIGURE 9. A suburban street redesigned to slow traffic and provide play space and pedestrian rest areas. From P. Bosselmann, "Redesigning Residential Streets," in A. V. Moudon (Ed.), *Public Streets for Public Use*, New York: Van Nostrand Reinhold, 1987. Reprinted by permission of the author.

residential district designated for pedestrian use through traffic bumps, bends, bollards, plantings, street furniture, and play equipment (Royal Dutch Touring Club, 1980). The positive impact of *woonerven* on both children's and teenager's street use has been documented (Eubank-Ahrens, 1987); but despite their European popularity, they have not been accepted in North America. Bosselmann (1987) has proposed variations of this concept that can be applied to North American streets of different traffic densities. Figure 9 shows one example.

In addition to street life and opportunities to observe diverse adult roles, a number of authors have argued that access to nature is a basic childhood need. Olds (1987) has argued that the sights, sounds, smells, and touch of nature offer infants and toddlers optimal levels of sensory diversity. In detailed ethnographic studies, Hart (1979) and R. Moore (1986) have recorded the importance of undeveloped land for school-age children and young adolescents, for whom it offers loose material for construction and fantasy play, physical challenge, and opportunities to become familiar with vegetation, small wild life, and natural cycles (Figures 10, 11). In a comparison of two Swedish housing estates, Bjorklid

FIGURE 10. World making with basic materials. A "waste corner" of a residential neighborhood that isn't going to waste. Picture courtesy of Roger Hart.

FIGURE 11. In Village Homes, Davis, California, surface water runoff flows through a system of ponds and channels, supporting wildlife, vegetation, and children. Picture courtesy of Robin C. Moore.

(1985) observed that an estate with varied wild topography that children could use as they pleased supported more play, and more diverse play, than an estate dominated by designated playgrounds. Not least in importance, preserving enclaves of nature serves children's long-term as well as short-term advantage: in addition to settings for play, it increases the ecological diversity and stability of the world they will inherit.

In opposition to children's need for nature stands planners' and designers' propensity to plan and design. Olwig (1986) has noted that there is a paradox in the need to "plan" for unplanned space. It requires that environmental professionals "de-construct" their typical view of nature as a resource that needs to be curbed and ordered, in order to leave children room in which to construct their own realities.

At its most diverse, site design offers the advantages of playgrounds, streets, and nature in close proximity. For this purpose, McGinty, Cohen, and Moore (1982; Figure 12) have recommended the

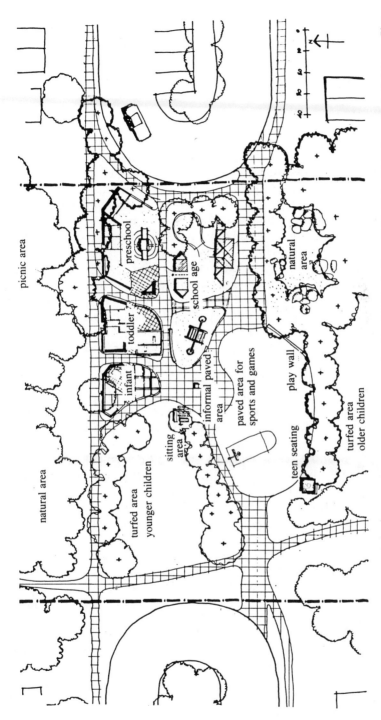

picnic area

natural area

preschool

school age

toddler

infant

sitting area

informal paved area

paved area for sports and games

play wall

teen seating

turfed area younger children

turfed area older children

natural area

N

10 10 20 30 40

FIGURE 12. Integrating a variety of neighborhood play areas (drawing by Tim McGinty). From T. McGinty, U. Cohen, & G. T. Moore, *Play Environments*, final report to the U.S. Department of the Army, Office of the Chief of Engineers, Washington, DC (TM 5-803-11), 1982. Reprinted by permission of the artist.

creation of play nodes throughout the residential environment where streets, playgrounds, and the natural environment meet.

OUTDOOR ACCESS FROM HIGH-RISE HOUSING

The relationship between dwelling height and children's use of the outdoors has been extensively researched. Since the 1960s, a succession of studies have claimed that because children living on high floors are more confined, they suffer from increased respiratory diseases, nervous disorders, boredom, aggression, delinquency, and family tensions, and decreased social skills, friendships, sense of control, school achievement, and opportunities for active play. After a careful review of this literature, Van Vliet (1983) has concluded that it has established that high rises complicate parental supervision and that preschool and school-age children spend less time outdoors when they live on high floors than when they live at or near ground level. He has noted that according to the model of outdoor access proposed by Hart (1978), high-rise housing pushes families into an "all-or-nothing" dilemma in which parents must either restrict children to the apartment, except for supervised outings, or relinquish protection when they allow them to leave alone. Van Vliet has cautioned, however, that this literature has failed to establish that other negative effects are the results of high-rise housing *per se:* there has been little agreement among studies regarding the definition of apartment housing; many studies have been anecdotal, or have failed to control for sociocultural variables; and there has been inadequate attention to the social, cultural, site, and design contexts of children's housing experience. In a similar vein, Churchman and Ginsberg (1981) have noted that this research has given little attention to important variables such as the number of families sharing building entrances, the form of access to dwellings, and the nature of on-site play opportunities.

In the light of their own review of housing evaluations, Cooper-Marcus and Hogue (1977) have recommended that families with children not be given housing above the third floor, or beyond the range of parents' sight and call, if it can be avoided. For parents with urban jobs, however, high-rise midcity housing can offer the benefit of less commuting time and greater family time. In this case, a report by Mackintosh (1985) has suggested how it can be made a viable housing form, given a relatively homogeneous stable population. Attending to residents' backgrounds and site and design features, as Van Vliet and Churchman and Ginsberg have recommended, Macintosh interviewed husbands and wives with at least one child between the ages of 2 and 10 at three well-regarded middle-income housing projects in New York, and at nearby high-rise towers with no site facilities. At two of the model projects (Stuyvesant Town and Peter Cooper Village), buildings clustered around internal pathways, streets, green space, and playgrounds. The third project (East Midtown Plaza; see Figure 13) combined a street-level playground with three terrace playgrounds with internal access only. More children at the model sites were allowed to play outside alone, at a younger age, than at the buildings without site facilities. Significantly more residents allowed their child out alone at the project with interior access playgrounds than at any other location; and a number of these children were under six years of age.

Another approach to ensuring children on high floors play space is to design corridors for play. Figure 14 presents some essential features of a convenient corridor.

Outdoor Access for Latchkey Children

Children left in self-care, or latchkey children, have become a familiar sign of the new social landscape as women have left home for the workplace in increasing numbers, without adequate child care support. What is less frequently acknowledged is that this phenomenon reflects changing out-of-school ecologies for all children. Recalling his own years growing up in the 1950s, Fink (1986) has noted that he and his brother spent very little time under his mother's direct supervision after school. That she and other mothers were at home meant that all children in the neighborhood came home after school, and that they moved freely in and out of each other's houses and across the neighborhood terrain in their play, secure in the proximity of adults if one should be needed. The result was that, "We got a total environment in which we felt connected to each other, to the physical resources of the neighborhood, and to the adults in the neighborhood" (p. 9). Without adult supervision, latchkey children may be at special risk with regard to security in the home and outdoors; but the latchkey phenomenon signifies the erosion of this total environment for all children.

Effects of the latchkey experience on children have been widely debated. When Robinson, Coleman, and Rowland (1986) analyzed major latchkey studies by location, they found environmental context to be a key factor in how well children adjusted to self-care. In urban studies, children left alone showed greater fear, lower school achievement, and poorer self-concepts and social relationships than adult-supervised peers. Rural and suburban studies showed no differences on any of these measures. Robinson, Coleman, and Rowland have noted that most

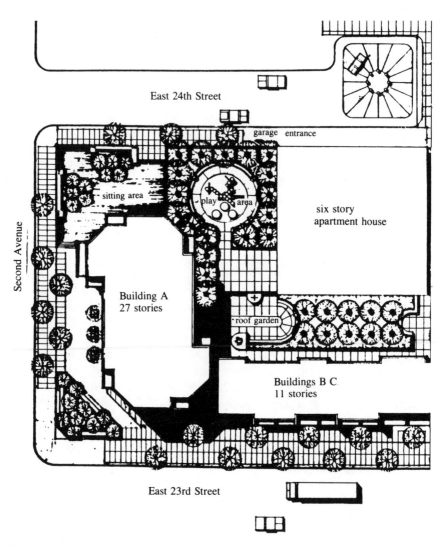

East 24th Street

garage entrance

sitting area

play area

six story apartment house

Second Avenue

Building A
27 stories

roof garden

Buildings B C
11 stories

East 23rd Street

FIGURE 13. One of three rooftop play areas in the East Midtown Plaza high-rise complex in New York. These second-floor-level playgrounds are accessible only through the building. A street-level playground is open to the public. An evaluation of residential satisfaction showed that children at East Midtown Plaza were much more likely to be allowed to play outside unsupervised than children at neighboring high-rise buildings with street-level playgrounds only or with no site facilities; and they were allowed outside alone at a younger age (an average age of six years). From E. Mackintosh, "Highrise Family Living in New York City," in E. L. Birch (Ed.), *The Unsheltered Woman*, New Brunswick: Center for Urban Policy Research, Rutgers University, 1985. Picture courtesy of Davis Brody & Associates, New York.

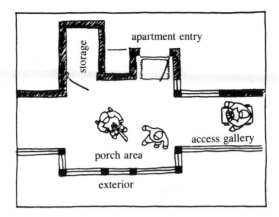

FIGURE 14. The basic components of an apartment "porch" in high-rise housing (drawing and design by Joel Shack). From L. C. Johnson, J. Shack, & K. Oster (Eds.), *Out of the Cellar and into the Parlour*, Ottawa: Canada Mortgage & Housing Corporation, 1980. Reprinted by permission of Canada Mortgage & Housing Corporation and the artist.

studies have confounded race, family income, and location. Nevertheless, they have argued that regardless of race or income, children in high-crime urban areas are more likely to be told to lock themselves in the dwelling alone and to fear intrusions, whereas children in nonurban areas face fewer restrictions on outdoor play.

One solution to this problem is school-age child care centers. At their best, they can coordinate challenging activities for all children in the community (Fink, 1986). Another approach is to strengthen social support for children both at home and outdoors. Steps in this direction are telephone help lines that children can call for comfort or advice, block parent programs in which adults at home volunteer to be available for emergency help, "grandparent" programs, supervised playgrounds, and check-in programs in which an adult at home contracts to provide after-school snacks, homework help, play arrangements, and community trips (Chawla, 1986). Together, these programs represent ways to free children from isolation in the home and to reconnect adults, children, and physical resources within the community.

CHILDREN'S PARTICIPATION IN DESIGN

In looking for housing, a primary consideration for parents may be finding a good place for their children; but as a rule children have little or no input into the shape of this place that is "for them." In contrast to an

extensive literature documenting children's participation in neighborhood planning and in classroom and playground design (reviewed
by Hart, 1987), very little has been published regarding their participation in the planning and design of housing. The explanation is probably
twofold. In this case, as in housing research in general, there is the
practical constraint that it is easier for professionals to work with children in public spaces than to intrude on the privacy of the home. Another hurdle is that, given limited space where family members must
coexist, parents as a rule exercise a territorial control that excludes children's participation. According to the family systems theorists Kantor
and Lehr (1975), the control of space, both physically and meta-
phorically, is *the* key issue in family functioning. Therefore, involving
children in housing design is not a simple matter of doing more: it
implies complex adjustments in family relationships.

Two practical factors must also be considered. One is that children
cannot be expected to have input into the form of their own home unless
their parents are empowered. A second consideration is that the creative
adaptation of space, once housing has been constructed, requires having space. Reminiscences of coming of age in large old houses contain
cherished memories of out-of-the-way corners: the attic, the garage, the
stair landing, spaces that could be colonized much like odd corners
outdoors (Cooper-Marcus, 1978). These niches have been largely elimi-
nated by the trend toward smaller dwellings. Children of the well-to-do
may still be assured their own bedrooms; but children in crowded
homes often have no place that they can call their own. To ease the
exigencies of economics, architects and interior designers need to give
more attention to recreating and evaluating small appropriable spaces
for children, such as window seats, alcoves, and lofts.

Within these constraints, a few initial steps have been taken to
involve children in the design and maintenance of housing. Bethany
(1981) has reported successful efforts by architects to engage children
ages five and older in the design of their own rooms by asking them to
select among alternative models. Baird and Lutkus (1982) have outlined
how architectural competence in childhood develops in step with spatial
cognition skills. Noschis (1982) has described how these abilities have
been demonstrated in children's full-scale modeling of future housing.
There are also numerous schemes of social architecture which have
already involved entire families in planning housing renovation and
construction (Hatch, 1984); unfortunately, the particular contributions of
children of different ages to these projects and the effects of participation
upon them have never been systematically documented.

One promising way to increase children's participation is to incorporate them into the design research process. Little (1983) has advocated personal project analysis, which can be applied to all stages of housing design and maintenance. To address researchers' need for quality-of-life indicators that reflect children's perspectives, the analysis elicits children's own subjective weightings of objects and events. Two components—a conjoint project matrix and a resource inventory—examine conflicts between children's activities and those of other family members, and identify resource availability and constraints. This information can then be applied to the goals of restructuring or relocating conflicting activities and providing needed resources. The analysis can be used on any scale from that of a room to that of a city. A limitation is that it is a self-report measure that requires a reflectiveness and articulateness that few children younger than 11 possess.

Another approach to involving children in housing issues is exemplified by the Children Creating Alternative Futures project described by Baldassari, Lehman, and Wolfe (1987). Implemented in three urban schools, the project engaged fifth and sixth graders in researching their neighborhoods and in envisioning change. The topics that the children investigated included housing quality, gentrification, and landlord–tenant relationships. Participants achieved observable gains in social and communication skills, creative problem solving, and self-esteem; but they were unable to effect anything but temporary environmental improvements because the adults who held power in their community had made no commitment to change.

The outcome of this project illustrates the rule that children's participation can be easily thwarted unless it is a legally mandated part of the planning process. One model of this mandate is the Norwegian Child Plan (Verktoykassa, 1983; described in Gaunt, 1987). To ensure that children's issues go on the municipal agenda for discussion and action, "the Child Plan consists of a situational report which describes assets, potentials, deficiencies, and problems from the point of view of children," along with "a programme for action which prescribes concrete measures and cost estimates as well as means for implementation" (Gaunt, 1987, p. 51). The plan can address any issue that impacts children's lives, such as safety standards, dwelling space standards, zoning, and access to child care and other neighborhood resources.

IMPLICATIONS FOR FUTURE RESEARCH AND APPLICATIONS

The preceding review of recent literature on children's and adolescents' housing has left some important facets of their experience and

needs unaddressed. In part, these omissions reflect the space restrictions of this chapter, which have dictated a focus on studies of
children's experience at the expense of larger social, economic, and political issues that impact it. In part, however, these omissions reflect
issues that have been neglected in the past. For research, design, and
policy to more fully reflect the realities of young people's lives, three
areas of work require concerted effort in the future: the integration of the
literature on children's housing experience with other relevant literatures; basic research on yet unexplored aspects of their experience; and
the more effective dissemination of existing knowledge.

RESEARCH INTEGRATION

Of fundamental importance, children's well-being must be identified with the well-being of the earth as a whole, whose condition they
will inherit. This identification must be made in practice as well as theory by integrating the literature on children's housing and site requirements with the literature on the construction of energy-efficient housing
and sustainable communities. It is noteworthy that conservation
schemes to promote urban forests and gardens (e.g., Spirn, 1984) or to
improve land and resource use (e.g., Van der Ryn & Calthorpe, 1986)
present opportunities to serve children's and adolescents' short-term as
well as long-term interests; but to be optimally served, these popula-
tions must be specifically addressed by these plans, and programs for
youth participation and ecological education must be incorporated.

Another urgent issue is child safety. Considering that many serious
and fatal accidents occur in the home, it is important to combine an
understanding of how they occur with design approaches to their prevention. Garling and Valsiner (1985) have collected recent theoretical
essays on parents' supervision of child safety. As these theories are
empirically tested, results will need to be integrated with earlier attempts to synthesize room-by-room guidelines for prevention (Jackson,
1977). Also of concern is indoor pollution through building materials
and the infiltration of outdoor air pollutants (Kane, 1985). Monitoring
building practices needs to be combined with efforts to define and promote ways in which parents can minimize children's exposure.

Research on parents' and children's environmental experience also
requires closer coordination. A burgeoning literature has begun to challenge traditional housing and planning practices on the grounds that
they have conformed to what practitioners thought women *should* want,
rather than what they in fact require (Peterson, 1987); but this work has
not yet been integrated with the literature on children's environments to

determine how places can best serve women, children, and adolescents simultaneously. The degree to which existing housing has served men's real needs also remains unclear (Peterson, 1987). For example, there have not yet been attempts to define environments that will make it easier for men to share parenting. A few model postoccupancy evaluations have attended to parents', children's, and adolescents' experiences at the same time (Cooper, 1975; Keller, 1981; Michelson, 1977; Popenoe, 1977; Sarkissian & Doherty, 1987). Work of this kind needs to become the rule rather than the exception.

NEW RESEARCH DIRECTIONS

The housing experiences of different family members need to be integrated in basic research as well as in literature syntheses. Little is yet known about how parents and children mediate each other's housing experience, or how housing mediates parent–child relationships. The following critical interactions have been barely explored.

1. Despite evidence that parents foster early childhood competence by effectively managing the availability of objects and floor space (White *et al.*, 1979), there has been little effort to simultaneously facilitate household management and indoor play through design. Innovations to this purpose should be evaluated for their effects upon parent–child interactions and children's freedom.

2. Despite evidence that women carry the main burden of housework even when they hold jobs outside the home (Peterson, 1987), there has been little design and evaluation of interiors with the goal of minimizing the time required for housekeeping and making it more easily shared by all family members. Attempts to simplify housework should be followed by attention to its effect on the quality of parent–child interactions.

3. Child maltreatment has been related to deteriorated housing, high residential instability, and low levels of residential satisfaction and neighboring (Bouchard, Beaudry, & Chamberland, 1985; Garbarino & Sherman, 1980; Sharp, 1984; Zuravin, 1987); but how specific housing features affect the dynamics of maltreatment remains unknown.

4. Along with a general neglect of the meaning of home among children, there has been particular neglect of the meaning of substandard housing and homelessness (for an exception, see Rivlin, 1986). Across North America and Western Europe, decreasing government subsidies of low-income housing and the loss of affordable units have pushed families with children into substandard housing, squatter hous-

ing, crowded housing, and—at the furthest extreme—onto the street (Huttman, 1985). In North America, this situation has been aggravated by discriminatory rental and zoning practices (Ritzdorf, 1986) and high poverty rates among families with children. The few studies of homeless children that exist (reviewed in Bassuk, Rubin, & Lauriat, 1986; Rivlin & Schwartzman, 1988) disclose a pattern of severely disrupted lives, along with developmental lags, depression, and anxiety. Reports on the effects of substandard housing and homelessness need to be balanced by evaluations of the effects of model intervention programs.

5. Families who manage to maintain their homes face rapidly rising housing costs. Little is known about the effects of rising costs upon family budget trade-offs, stress, and interpersonal dynamics.

6. On the positive side, recent attention has been given to reinvolving residents in the design and construction of their own homes through techniques of participatory architecture (Hatch, 1984). An intriguing subject for the future is the effect of families' involvement in this process upon children's environmental attitudes and behavior.

As a general rule, inconclusive findings in some areas have shown that future studies will need to control for the joint effects of gender, age, family income, housing form, and site resources in evaluating children's experience. With attention to all of these variables, children's own assessments of their environment need to be compared with adult assessments, and given serious consideration.

RESEARCH DISSEMINATION

It is as important to understand why design and policy recommendations are disregarded or implemented as it is to determine what they should be. To this end, case studies of development, design, and management processes that thwart innovation need to be compared to studies of successful attempts to overcome obstacles to child-oriented decision making. One key to implementation is the effective dissemination of research results.

As this chapter has shown, some family housing uses and preferences have been well documented; yet standard building practices remain at variance with research findings. This chapter has cited numerous studies that have disclosed that children make active use of a bedroom of their own, or a room large enough to share comfortably with a sibling, and that this personal space is highly prized. Traditional practices, however, allot minimal bedroom space to children in favor of a master bedroom for parents, and expect same-sex siblings to double up. Another discrepancy is that young children have been observed to gravitate to the kitchen, as the center of family activity, but this room is rarely designed for their use. As a general rule, both indoors and outdoors, children's need for a space of their own and space for active play is given low priority. Outdoors, researchers have stressed the importance of undefined, appropriable spaces where children can create their own imaginary worlds and spontaneous architecture, but these "waste" places continue to be designed away.

In part, this disregard for children reflects developers' tendency to market housing to parents by emphasizing adult privileges, even though resulting housing choices may not in fact reflect parents' true preferences. In part, constraints on play reflect high insurance premiums and fear of accident liability. There is a special need, therefore, to distribute information on developmental needs, parent and child preferences, and accident rates under different situations among developers, insurers, and building managers.

There is also a need to better promote advances among research and design professionals. A frustrating limitation of the current field is that continental European research on children's environments remains largely inaccessible to the English-speaking world. More effort needs to be given to the translation of this material.

Accepting that social change is ongoing, research, programming, design, construction, evaluation, and research dissemination must be seen as a continuous cycle. As new attempts at rezoning, financing, and designing for diverse family types are tried, successful solutions need to be publicized. To ensure that children will be attended to, special recognition needs to be given to successful plans that formally incorporate them into the housing and site planning process. As housing costs, shortages of affordable housing, and homelessness have increased, so has public concern that housing must be acknowledged to be a basic human right. For children, the construction of good housing is not only a basic right. It is solid affirmation that preceding generations have taken care to make the present a secure home for the unfolding of the future.

ACKNOWLEDGMENTS

The author is indebted to Christine Bevington for a close reading of an earlier draft of this chapter, and to her, Roger Hart, and Robin Moore for their sharing of resource materials. Each stage of the writing has benefited from Gary Moore's deft, constructive editing. Evalyn Verhey and Jayna Oakley have helped prepare the illustrations for publication. A grant from the Kentucky State University Faculty Research Fund covered expenses incurred during writing.

REFERENCES

Aiello, J. R., Thompson, D. E., & Baum, A. (1985). Children, crowding, and control: Effects of environmental stress on social behavior. In J. F. Wohlwill & W. Van Vliet (Eds.), *Habitats for children*. Hillsdale, NJ: Erlbaum.

Baird, J. C., & Lutkus, A. D. (1982). From spatial perception to architectural construction. In J. C. Baird & A. D. Lutkus (Eds.), *Mind, child, architecture*. Hanover, NH: University Press of New England.

Baldassari, C., Lehman, S., & Wolfe, M. (1987). Imaging and creating alternative environments with children. In C. Weinstein & T. G. David (Eds.), *Spaces for children* (pp. 241–268). New York: Plenum.

Barbey, G. F. (1974). Anthropological analysis of the home concept. In D. H. Carson (Ed.), *Man–environment interactions* (Vol. 3, pp. 143–149). Stroudsburg, PA: Dowden, Hutchinson & Ross.

Barry, V. T. R. (1982). Kidspace: Family life in the city. *Children Today, 11,* 11–15.

Bassuk, E. L., Rubin, L., & Lauriat, A. S. (1986). Characteristics of sheltered homeless families. *American Journal of Public Health, 76*(9), 1097–1101.

Berg, M., & Medrich, E. A. (1980). Children in four neighborhoods. *Environment and Behavior, 12*(3), 320–348.

Bethany, M. (1981). What children want. In C. Donovan (Ed.), *Living well*. New York: Quadrangle/New York Times.

Bevington, C. B. (1987a, May). A quilt-plan. *Interior Design*, pp. 326–327.

Bevington, C. B. (1987b). Housing the homeless mother and child. *Women & Environments, 10*(1), 16–17.

Bjorklid, P. (1985). Children's outdoor environment from the perspectives of environmental and developmental psychology. In T. Garling & V. Valsiner (Eds.), *Children within environments* (pp. 91–106). New York: Plenum.

Bosselmann, P. (1987). Redesigning residential streets. In A. V. Moudon (Ed.), *Public streets for public use* (pp. 321–330). New York: Van Nostrand Reinhold.

Bouchard, C., Beaudry, J., & Chamberland, C. (1985). *An ecological approach to child maltreatment: Relationship between residential satisfaction and the incidence of child abuse.* (Mimeo, Laboratoire de recherche en écologie humaine et sociale, Université du Québec a Montréal.)

Bronfenbrenner, U. (1979). *The ecology of human development.* Cambridge, MA: Harvard University Press.

Caldwell, B. (1968). Inventory of home stimulation. Little Rock, AK: Center for Early Development & Education.

Canada Mortgage & Housing Corporation. (1978). *Play spaces for preschoolers.* Ottawa: Author.

Canada Mortgage & Housing Corporation. (1979). *Play opportunities for school-age children, 6 to 14 years of age.* Ottawa: Author.

Chawla, L. (Ed.). (1986). Latchkey children in their communities [Special issue]. *Children's Environments Quarterly, 3*(2).

Churchman, A. (1980). Children in urban environments: The Israeli experience. *Ekistics, 47*(281), 105–109.

Churchman, A., & Ginsberg, Y. (1981). *Housing type and children's outdoor play—Is there a relationship?* Paper presented at the Eighth World Conference of the International Playground Association, Rotterdam.

Cohen, U., Hill, A. B., Lane, C. G., McGinty, T., & Moore, G. T. (1979). *Recommendations for child play areas*. Milwaukee: University of Wisconsin-Milwaukee, Center for Architecture and Urban Planning Research.

Cooper, C. C. (1975). *Easter Hill Village*. New York: Free Press.

Cooper-Marcus, C. (1974). Children's play behavior in a low-rise, inner-city housing development. In D. H. Carson (Ed.), *Man-environment interactions* (Vol. 3, pp. 197–211). Stroudsburg, PA: Dowden, Hutchinson & Ross.

Cooper-Marcus, C. (1978). Remembrance of landscapes past. *Landscape*, 22(3), 34–43.

Cooper-Marcus, C., & Hogue, L. (1977). Design guidelines for high-rise family housing. In D. Conway (Ed.), *Human response to tall buildings* (pp. 240–277). Stroudsburg, PA: Dowden, Hutchinson & Ross.

Cooper-Marcus, C., & Sarkissian, W. (1986). *Housing as if people mattered*. Berkeley: University of California Press.

Csikszentmihalyi, M., & Rochberg-Halton, E. (1981). *The meaning of things*. Cambridge: Cambridge University Press.

DeJoy, D. M. (1983). Environmental noise and children: Review of recent findings. *Journal of Auditory Research*, 23, 181–194.

Eubank-Ahrens, B. (1987). A close look at the users of *woonerven*. In A. V. Moudon (Ed.), *Public streets for public use* (pp. 63–79). New York: Van Nostrand Reinhold.

Filipovitch, A. J., Juliar, K., & Ross, K. D. (1981). Children's drawings of their home environment. In A. E. Osterberg, C. P. Tiernan, & R. A. Findlay (Eds.), *Design research interactions* (pp. 258–264). Washington, DC: Environmental Design Research Association.

Fink, D. (1986). School-age child care: Where the spirit of neighborhood lives. *Children's Environments Quarterly*, 3(2), 9–12.

Flade, A. (1986). Evaluation of housing floor plans with regard to meeting family needs. *Children's Environments Quarterly*, 3(1), 68–72.

Francis, M. (1985). Children's use of open space in Village Homes. *Children's Environments Quarterly*, 1(4), 36–38.

Franck, K. A. (1985). New households, old houses: Designing for changing needs. *Ekistics*, 52(310), 22–27.

Franck, K. A., & Ahrentzen, S. (Eds.). (1989). *New households, new housing*. New York: Van Nostrand Rheinhold.

Garbarino, J., & Sherman, D. (1980). High-risk families and high-risk neighborhoods. *Child Development*, 51, 188–198.

Garling, T. & Valsiner, V. (Eds.). (1985). *Children within environments*. New York: Plenum.

Gaunt, L. (1980). Can children play at home? In P. F. Wilkinson (Ed.), *Innovation in play environments*. New York: St. Martin's Press.

Gaunt, L. (1987). Room to grow—in creative environments or on adult premises? *Scandinavian Housing and Planning Research*, 4, 39–53.

Greenman, J. (1987). *Caring spaces, learning places*. Redmond, WA: Exchange Press.

Hart, R. (1978). Children's exploration of tomorrow's environments. *Ekistics*, 45(272), 387–390.

Hart, R. (1979). *Children's experience of place*. New York: Irvington.

Hart, R. (1987). Children's participation in planning and design: Theory, research, and practice. In C. S. Weinstein & T. G. David (Eds.), *Spaces for children* (pp. 217–239). New York: Plenum.

Harvey, J. (1982). *Vandalism in the residential environment*. Ottawa: Canada Mortgage & Housing Corporation.

Hatch, C. R. (1984). *The scope of social architecture*. New York: Van Nostrand Reinhold.

Hayden, D. (1982). *The grand domestic revolution*. Cambridge, MA: MIT Press.

Hayden, D. (1984). *Redesigning the American dream*. New York: Norton.

Hayward, D. G. (1975). Home as an environmental and psychological concept. *Landscape*, 20(1), 2–9.

Held, R., & Hein, A. (1963). Movement-produced stimulation in the development of visually guided behavior. *Journal of Comparative and Physiological Psychology, 56*, 872–876.

Homel, R., & Burns, A. (1985). Through a child's eye: Quality of neighborhood and quality of life. In I. Burnley & J. Forrest (Eds.), *Living in cities* (pp. 103–115). Sydney: Allen & Unwin.

Huttman, E. D. (1985). Transnational housing policies. In I. Altman & C. Werner (Eds.), *Home environments*. New York: Plenum.

Jackson, R. H. (Ed.). (1977). *Children, the environment, and accidents*. Kent: Pitman Medical Publishing.

Johnson, L. C. (1987). The developmental implications of home environments. In C. S. Weinstein & T. G. David (Eds.), *Spaces for children* (pp. 139–157). New York: Plenum.

Johnson, L., Shack, J., & Oster, K. (1980). *Out of the cellar and into the parlour*. Ottawa: Canada Mortgage and Housing Corporation.

Kane, D. N. (Ed.). (1985). *Environmental hazards to young children*. Phoenix: Oryx Press.

Kantor, D., & Lehr, W. (1975). *Inside the family*. San Francisco: Jossey-Bass.

Keller, S. (1981). Women and children in a planned community. In S. Keller (Ed.), *Building for women* (pp. 67–75). Lexington, MA: Heath.

Ladd, F. C. (1972). Black youths view their environments: Some views of housing. *American Institute of Planners Journal, 38*, 108–116.

Little, B. R. (1983). Personal projects. *Environment and Behavior, 15*(3), 273–309.

Lynch, K. (Ed.). (1977). *Growing up in cities*. Cambridge, MA: MIT Press.

McGrath, M., & McGrath, N. (1978). *Children's spaces*. New York: Morrow.

McGinty, T., Cohen, U., & Moore, G. T. (1982). *Play environments*. Final report to the U.S. Dept. of the Army, Office of the Chief of Engineers, Washington, DC (TM 5-803-11).

Mackintosh, E. (1985). Highrise family living in New York City. In E. L. Birch (Ed.), *The unsheltered woman* (pp. 101–119). New Brunswick, NJ: Rutgers University, Center for Urban Policy Research.

Michelson, W. (1968). *The physical environment as a mediating factor in school achievement*. Paper presented at the annual meeting of the Canadian Sociology and Anthropology Association, Calgary.

Michelson, W. (1977). *Environmental choice, human behavior, and residential satisfaction*. New York: Oxford University Press.

Moore, G. T. (1982). *Methodological considerations in field studies of developmental psychology and the built environment*. Paper presented at the annual meeting of the Environmental Design Research Association, Washington, DC.

Moore, G. T. (1985). State of the art in play environment research and applications. In J. L. Frost & S. Sunderlin (Eds.), *When children play* (pp. 171–192). Wheaton, MD: Association for Childhood Education International.

Moore, G. T., Lane, C. G., Hill, A. B., Cohen, U., & McGinty, T. (1979). *Recommendations for child care centers*. Milwaukee: University of Wisconsin-Milwaukee, Center for Architecture and Urban Planning Research.

Moore, R. C. (1986). *Childhood's domain*. London: Croom Helm.

Moore, R. C. (1987). Streets as playgrounds. In A. V. Moudon (Ed.), *Public streets for public use* (pp. 45–62). New York: Van Nostrand Reinhold.

Moore, R. C., & Young, D. (1978). Childhood outdoors: Toward a social ecology of the landscape. In I. Altman & J. F. Wohlwill (Eds.), *Children and the environment* (pp. 83–130). New York: Plenum.

Neperud, R. W. (1975). Favorite places. *Journal of Environmental Education, 6,* 27–31.

Noschis, K. (1982). The child in the laboratory. In J. C. Baird & A. D. Lutkus (Eds.), *Mind child architecture.* Hanover, NH: University Press of New England.

Olds, A. R. (1987). Designing settings for infants and toddlers. In C. S. Weinstein & T. G. David (Eds.), *Spaces for children* (pp. 117–138). New York: Plenum.

Olwig, K. R. (1986). The childhood "deconstruction" of nature and the construction of "natural" housing environments for children. *Scandinavian Housing and Planning Research, 3,* 129–143.

Parke, R. D. (1978). Children's home environments: Social and cognitive effects. In I. Altman & J. F. Wohlwill (Eds.), *Children and the environment* (pp. 34–81). New York: Plenum.

Parke, R. D., & Sawin, D. B. (1979). Children's privacy in the home: Developmental, ecological and child-rearing determinants. *Environment and Behavior, 11,* 87–104.

Peterson, R. B. (1987). Gender issues in the home and urban environment. In E. H. Zube & G. T. Moore (Eds.), *Advances in environment, behavior, and design* (Vol. 1, pp. 187–218). New York: Plenum.

Piaget, J. (1952). *The origins of intelligence in children* (M. Cook, Trans.). New York: International Universities Press.

Pollowy, A.-M. (1977). *The urban nest.* Stroudsburg, PA: Dowden, Hutchinson & Ross.

Popenoe, D. (1977). *The suburban environment.* Chicago: University of Chicago Press.

Ritzdorf, M. (1986). Adults only: Children and American city planning. *Children's Environments Quarterly, 3*(4), 26–33.

Ritzdorf, M. (1987). Planning and the intergenerational community. *Journal of Urban Affairs, 9*(1), 79–89.

Rivlin, L. (1986). A new look at the homeless. *Social Policy, 16,* 3–10.

Rivlin, L., & Schwartzman, J. (1988). Street children and homeless children [Special issue]. *Children's Environments Quarterly, 5*(1).

Robinson, B. E., Coleman, M., & Rowland, B. H. (1986). The after-school ecologies of latchkey children. *Children's Environments Quarterly, 3*(2), 4–8.

Royal Dutch Touring Club. (1980). *Woonerf.* The Hague: Royal Dutch Touring Club.

Rubenstein, J. L., & Howes, C. (1979). Caregiving and infant behavior in day care and in homes. *Developmental Psychology, 15,* 1–24.

Saegert, S., & Hart, R. (1978). The development of sex differences in the environmental competence of children. In M. Salter (Ed.), *Play.* Cornwall, NY: Leisure Press.

Sarkissian, W., & Doherty, T. (1987). *Living in public housing.* Red Hill, A.C.T.: Royal Australian Institute of Architects Education Division.

Schiavo, R. S. (1988). Age differences in assessment and use of a suburban neighborhood among children and adolescents. *Children's Environments Quarterly, 5*(2), 4–9.

Schiavo, R. S. (1987). Home use evaluation by suburban youth: Gender differences. *Children's Environments Quarterly, 4*(4), 8–12.

Sebba, R., & Churchman, A. (1983). Territories and territoriality in the home. *Environment and Behavior, 15*(2), 191–210.

Shack, J. (1987). Evaluation of a housing demonstration project designed for play and child care. In J. Harvey & D. Henning (Eds.), *Public environments* (pp. 32–41). Washington, DC: Environmental Design Research Association.

Sharp, C. (1984). Environmental design and child maltreatment. In D. Duerk & D. Campbell (Eds.), *The challenge of diversity* (pp. 66–73). Washington, DC: Environmental Design Research Association.

Spirn, A. (1984). *The granite garden.* New York: Basic Books.

Sweaney, A. L., Inman, M. A., Wallinga, C. R., & Dias, S. (1986). The perceptions of preschool children and their families' social climate in relation to household crowding. *Children's Environments Quarterly, 3*(4), 10–15.

Thornberg, J. M. (1974). Children's conception of places to live in. In D. H. Carson (Ed.), *Man–environment interactions* (Vol. 3, pp. 178–190). Stroudsburg, PA: Dowden, Hutchinson & Ross.

Van der Ryn, S., & Calthorpe, P. (Eds.). (1986). *Sustainable communities.* San Francisco: Sierra Club Books.

Van Vliet, W. (1983). Families in apartment buildings: Sad storeys for children? *Environment and Behavior, 15,* 211–234.

Van Vliet, W. (1985). The methodological and conceptual basis of environmental policies for children. *Prevention in Human Services, 4*(1/2), 59–78.

Verktoykassa (1983). Redskap for planlegging av barns naermiljo [Instrument for planning of children's environment]. Oslo: Norwegian Institute for Urban and Regional Research.

Verwer, D. (1980). Planning residential environments according to their real use by children and adults. *Ekistics, 47*(281), 109–113.

Wachs, T. D. (1979). Proximal experience and early cognitive-intellectual development: The physical environment. *Merrill-Palmer Quarterly, 25,* 3–41.

Wachs, T. D., & Gruen, G. E. (1982). *Early experience and human development.* New York: Plenum.

Wattenberg, B. J. (1987). *Baby boom to birth dearth.* New York: Pharos.

Weinstein, C. S., & David, T. G. (Eds.). (1987). *Spaces for children.* New York: Plenum.

White, B. L., Kaban, B., & Attanucci, J. (1979). *The origins of human competence.* Lexington, MA: Heath.

Wilson, S. (1980). Vandalism and "defensible space" on London housing estates. In R. V. G. Clarke & P. Mayhew (Eds.), *Designing out crime.* London: Her Majesty's Stationery Office.

Wohlwill, J. F. (1980). The confluence of environmental and developmental psychology: Signpost to an ecology of development? *Human Development, 23,* 354–358.

Wohlwill, J. F., & Heft, H. (1987). The physical environment and the development of the child. In D. Stokols & I. Altman (Eds.), *Handbook of environmental psychology* (Vol. 1, pp. 281–328). New York: Wiley.

Wohlwill, J. F., & Van Vliet, W. (Eds.). (1985). *Habitats for children.* Hillsdale, NJ: Erlbaum.

Wolfe, M. (1978). Childhood and privacy. In I. Altman & J. F. Wohlwill (Eds.), *Children and the environment* (pp. 175–222). New York: Plenum.

Zeisel, J., & Welch, P. (1981). *Housing designed for families.* Cambridge, MA: Joint Center for Urban Studies of MIT & Harvard.

Ziegler, S., & Andrews, H. F. (1987). Children and built environments: A review of methods for environmental research and design. In R. Bechtel, R. Marans, & W. Michelson (Eds.), *Methods in environmental and behavioral research* (pp. 301–336). New York: Van Nostrand Reinhold.

Zinn, H. (1980). The influence of home environments on the socialization of children. *Ekistics, 47*(281), 98–102.

Zuravin, S. (1987). The ecology of child maltreatment: Identifying and characterizing high-risk neighborhoods. *Child Welfare, 66,* 497–506.

IV

ADVANCES IN SOCIOBEHAVIORAL RESEARCH

7

Environmental Meaning

MARTIN KRAMPEN

TWO APPROACHES TO THE STUDY OF
ENVIRONMENTAL MEANING

In this chapter, advances in research on environmental meaning are addressed by connecting the present state of the art in research to the past, and by extrapolating from the present to possible future developments. Two concurrent approaches to the study of environmental meanings are addressed: semiotics and environmental psychology. The chapter concludes with a discussion of the ecological approach and its potential contribution to understanding environmental meaning.

The two approaches stem from different origins and traditions. Both use some similar terms; however, they have different definitions. For example, in semiotics, at least in the tradition of Peirce, a material *sign* may be related to the object it stands for in three ways. One relationship, called *iconic*, occurs if the sign has properties in common with the object it stands for (as in the case of a portrait). In another relationship, called *indexical*, a sign is an index if it functions by temporal or spatial contact with the object (as in a weather vane). In the third relationship, called *symbolic*, a sign is based on a conventional relationship to its object (such as in a license plate of a car). To appreciate the differences of the two

Martin Krampen • University of the Arts Berlin, Federal Republic of Germany; Am Hochstrasse 18, D-7900 Ulm-Donau, Federal Republic of Germany.

approaches, it is useful to compare the above characterization of the sign in semiotics with the distinction that is sometimes made in philosophy or in social sciences, including environmental psychology, between *signs* (having univocal meaning) and *symbols* (having multivocal meaning), or the attribution of "symbols" to "high-level meaning."

The semiotic approach to environmental meaning is sometimes confused with a linguistic approach. Architectural and environmental semiotics are guilty of this confusion because they tried to emulate linguistic models. But the semiotic study of meaning is not confined to verbal meaning and, therefore, investigates other than verbal signs such as are present in natural and human environments. On the other hand, the term "nonverbal communication," proposed by Rapoport (1982) as the most promising way to studying environmental meaning, suffers from its relationship to verbal communication, with which it is held to be complementary, at least in linguistics (e.g., Posner, 1986). Its relationship to communication in general proves to be problematic as long as communication is thought of in terms of an intentional act of a sender. This seems not to be applicable to the natural environment and, similarly, there are large parts of material culture where no senders with intentions to communicate can be found.

Another problem of comparing the semiotic and environmental psychology approaches results from differences in the meaning of meaning. For semioticians, *meaning* is a property of signs, based on the interpretation of a sign's signifier (its material body) in relation to its signified (or meaning). This meaning of meaning is very technical and narrow. To the social scientist the term *meaning* seems very broad and in need of further specification. Thus Rapoport (1982) proposes to differentiate between different "levels" of meaning (e.g., a lower instrumental, an intermediate sociological, and a higher cosmological and "symbolic" level).

Finally, the aim of the two approaches to environmental meaning seems to be different. Whereas the semiotic approach favors the study of synchronic semiotic structures, that is, of sign systems or codes in the natural or human environment, the approach of environmental psychology is mainly concerned with environment–behavior relationships. This difference in aims is reflected in the methods applied. The methods of semiotics are often derived from linguistics, as is the case of the commutation test, where the substitution of one sign by another should precipitate a change of meaning (cf. *bed* vs. *red*). Those of environmental psychology stem from the tradition of empirical and quantitative research.

These differences need not, however, lead to contradictory results.

Rather, it appears to be promising to treat the results of semiotic investigations as hypotheses that may be tested by empirical research.

In this chapter, the origins and current state of the art of each approach are presented separately and then contrasted with a third approach—the ecological—in order to draw conclusions for further research and applications to environmental problem solving.

THE ORIGINS OF THE SEMIOTIC APPROACH TO RESEARCH ON ENVIRONMENTAL MEANING

From its beginnings in antiquity, semiotics has always dealt with practical problems. Well known is the example of Greek medicine, where semiotics was the science of symptoms. The Romans used signs such as the flight of birds to predict future events. In medieval times, heraldic signs were used to identify the bearer as friend or foe. In the age of enlightenment, the aesthetic merits of verbal and visual signs were compared. And in our time of international traffic, pictographs are used to communicate across language barriers. It is only since the time of enlightenment that applied semiotics has been flanked by comparative semiotics studying different sign types and their classification. Theoretical semiotics finally developed in the last century.

Advocates of theoretical semiotics have had a hard time convincing the established sciences that, as a metascience, it was a useful heuristic to unifying all scientific approaches. Thus in the present century, semiotics began to invade a number of subject matters for which a proper scientific approach was not yet established. This led to a variety of hyphenated fields of semiotics such as film-semiotics, theatre-semiotics, and architectural-semiotics. The oldest example of a semiotic treatment of architecture is found in Jan Mukarovsky's 1937–1938 article "On the Problem of Function in Architecture" (Mukarovsky, 1978) in which a multifunctional model of architectural meaning was postulated. But the "new wave" of semiotics of architecture goes back to the post World War II 1950s in Italy where a massive building boom in the big cities began to deface the urban image. Architectural semiotics was launched from two sources: by architects, either practitioners or historians, in search of a theory of architecture and/or environmental meaning (de Fusco, 1967; Gamberini, 1953, 1959, 1961; Koenig, 1964, 1970; Scalvini, 1968, 1975), and by philosophers and semioticians applying semiotics to urbanistic or architectural practice (Eco, 1968, 1972; Garroni, 1964, 1972).

This trend was soon taken over by British authors (Broadbent, 1969,

1975; Jencks, 1969), and by architects in France (Castex & Panerai, 1979; Hammad, 1979; Ostrowetsky & Bordreuil, 1979; Renier, 1979), where they formed a school around the semiotician Greimas (1974). At the first Congress of the International Association for Semiotic Studies (IASS) in Milan in 1974, a survey on current work on the semiotics of architecture was given by Krampen (1979b), who quoted more than 100 pertinent sources and studies in architecture.

The volume of semiotic research on environmental and architectural meaning has grown considerably since then. To consider single publications in various professional and other journals is impossible. Fortunately a recent *Encyclopedic Dictionary of Semiotics* (Sebeok, 1986) and a *Handbook of Semiotics* (Nöth, 1985) are available, which collate the information on architectural semiotics, on the semiotics of settlement space, and on the semiotics of urban culture (see also Gottdiener, 1986; Lagopoulos, 1986; Preziosi, 1986). Moreover, the proceedings of the 1974 Milan, the 1979 Vienna and the 1984 Palermo Conferences of the IASS contain contributions by the international "hard core" of researchers in environmental meaning. In addition, two important books by Preziosi (1979a, 1979b) have characterized the present state of the art in semiotic research on environmental meaning: *The Semiotics of the Built Environment* and *Architecture, Language, and Meaning.* Some single contributions on the topic can also be found in the various conference proceedings of the Environmental Design Research Association and the International Association for the Study of People and Their Physical Surroundings.

ENVIRONMENTAL MEANING IN SEMIOTICS

The beginning of the semiotic study of architectural and spatial meaning was marked by the efforts of different authors to mold the problems presented by these fields to fit models proposed by the various classics of semiotics and linguistics. Such classic authorities were de Saussure (in de Fusco, 1960), Morris (in Koenig, 1964, 1970), Peirce (in Blomeyer & Helmholtz, 1976; Broadbent, 1980d; Walther, 1974), Hjelmslev (in Greimas, 1974), and Chomsky (in Boudon, 1973, 1974; Broadbent, 1980a).

If proliferation of a semiotic model's application is a measure of its success, the most successful was that of the Danish linguist Hjelmslev (1968–1971). In his model he makes a distinction (first proposed by de Saussure) between the *form* and the *substance* of both a sign's signifier (i.e., material body) and signified (i.e., meaning). In Hjelmslev's terms, the distinction is between *form* and *substance* on the levels of expression

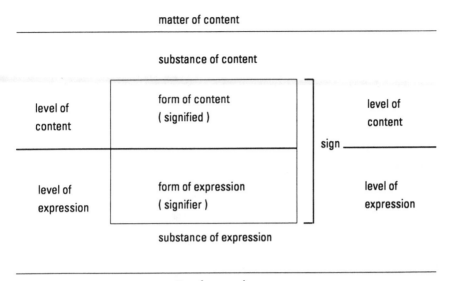

matter of content

substance of content

| level of content | form of content (signified) | | level of content |
| level of expression | form of expression (signifier) | | level of expression |

sign

substance of expression

matter of expression

FIGURE 1. Hjelmslev's model of a sign: A sign is carved out of the (formless) matters of expression and content. The sign results from the projection of a formal "grid" on both matters, thus resulting in a formed substance on both levels, that of expression and that of content. The sign proper is the unit made up of the form of expression (signifier) and the corresponding form of content (signified).

and content of a sign (Figure 1). One sign's combined form of expression and content may then become the form of expression of another sign with a secondary level of content attached to it (Figure 2). This type of composite sign, called *connotative* by Hjelmslev, played an important role in the analyses of Barthes (1964). The work of many researchers in the fields of architectural and spatial meaning consisted of identifying the substance from which a spatial form was derived and looking for its coordinated form of content and the substance from which it had been carved out. Following the model of linguistic research for minimal units of language (i.e., phonemes), this led to a search for the minimal formal units of expression and their correlated minimal forms of content in architectural or spatial problems. Likewise it led to the distinction of primary (denotative) and secondary (connotative) functions in the semiotics of architecture and space. *Primary* functions were said to be more material (e.g., shelter), and *secondary* ones to be more symbolic (e.g., the religious function of a spire). An interesting example of this approach is Eco's (1968) early attempt to formulate an architectural semiotics.

Another key term of semiotics is *code*, referring to entire systems of

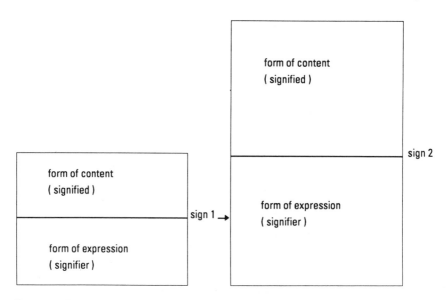

FIGURE 2. Denotative and connotative signs: According to Hjelmslev, a denotative sign consisting of a form of expression (signifier) and a form of content (signified) may become the form of expression (signifier) of a second sign with its own form of content (signified). This second sign is called connotative.

signs. Eco claimed that architectural codes (i.e., sign systems) based only on architectural forms and their functional content had a limited scope of application to existing examples of buildings. Therefore, a broader analysis of architecture and spatial meaning should be built on architectural and extra-architectural codes. One code of the latter variety, according to Eco (1968), is the code of proxemics (Hall, 1966), in which interpersonal distances reflect different functions, such as seeing a person's face in detail or perceiving the whole figure of a person. Distance may serve as the signifier of a connotative sign, the significance of which is the regulation of social distance in addition to mere person perception. Architectural or design elements, such as a table top of a certain size, not only have the simple function of accommodating objects at a certain height, but may also be used to regulate the distance between two persons, such as an employer and an employee. In this way, the combination between an architectural signifier (in this case the table top of a certain size) and a proxemic signified (a distance with certain consequences for personal perception) becomes in turn the signifier of a sign, which has an additional secondary signified (distinction of social or organizational levels; Figure 3).

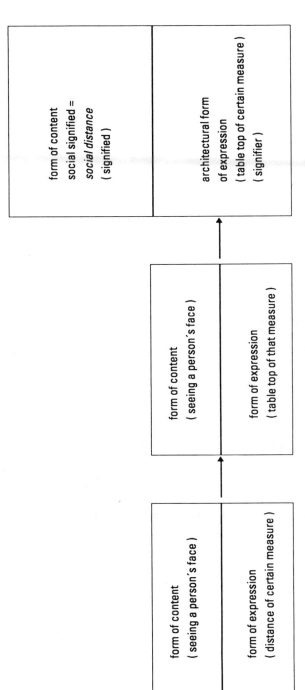

FIGURE 3. In a (denotative) proxemic sign made of an interpersonal distance as the signifier and the resulting person perception as a signified, the signifier (distance) may be substituted by an architectural object incorporating that distance (e.g., a table top). The resulting (denotative) architectural sign may become in turn the signifier of an (connotative) architectural sign regulating social distance.

Eco (1980) and Jencks (1980) applied this type of analysis to the architectural sign "column," which does not have the simple function of sustaining the architrave, but may, together with this function, become the signifier of another sign with the meaning "classical tradition." Scalvini (1980) used this multiple-level analysis to separate "tectonics" on the denotative (i.e., functional) level from "architecture" on the connotative level. This distinction gave rise to prolonged polemics against the "aesthetic fallacy" in architectural semiotics (Eco, 1968).

THE CONTRIBUTION OF GREIMAS

By far the most extensive use of Hjelmslev's model has been made by Greimas (1974, 1979) and his school (Fauque, 1979; Hammad, 1979). For Greimas (1979), every "extension," such as an earthly surface, is a substance in the sense of Hjelmslev's distinction between substance and form. If this *substance* "extension" is touched by human action, it becomes "space," that is, a *form* carved out of that substance.

Although the spatial signifier is coextensive with the natural world, it can be defined as a signifier-space on the "level of expression" (cf. Figure 1 above) only if it is coordinated with a specific signifying-form on the "level of content." This form on the level of content is carved out of a substance called "culture" by Greimas. A spatial sign consists thus of two levels. On the level of expression it is a space-form created by human intervention with spatial extension. On the level of content it is a cultural form delimited within culture (Figure 4).

A general semiotics of space (Greimas calls it a "general topological semiotics") does not yet exist. It has to be derived by constructing many particular semiotics such as, for instance, a semiotics of the city. A "topological" object such as the city is constituted by properties common to all of its parts. It is thus a multifaceted affair. Equally as multifaceted are the cultural forms of content connected to these parts. Consequently such an object can only be asserted by recurring themes of possible "readings."

According to Greimas, the level of content, the ideology of today's city, is organized on three polar thematic axes: the aesthetic (beautiful/ugly), the political (social and moral soundness/social and moral unhealthiness), and the rational (effective functioning/not functioning). This multilevel reading is further complicated by the opposition "society/individual" because the three axes may have as their subject matter either urban culture as a whole or the lifestyle of a single citizen. This model of the city's level of content is not only apt for "reading" the city but also provides a deep structure for generating beautiful or ugly, social

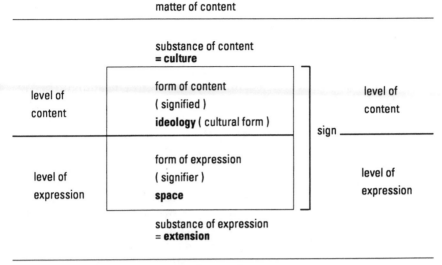

matter of content

substance of content
= **culture**

level of content	form of content (signified) **ideology** (cultural form)	level of content
		sign
level of expression	form of expression (signifier) **space**	level of expression

substance of expression
= **extension**

matter of expression

FIGURE 4. Application of Hjelmslev's model to "topological semiotics" by Greimas. Space is a special form of expression resulting from the formed substance of extension; ideology is a special form of content within the substance of culture.

or asocial, and functional or dysfunctional cities. Since the categories are formal, they are open to variable semantic interpretations depending on different cultures. For example, the olfactory and auditory stimuli of an Oriental city may be considered pleasant by the inhabitants and unpleasant by a Westerner. The compatibility of the positive or negative aspects of community culture and individual lifestyles are regulated by certain rules that determine the priorities between the aesthetic, political, and functional axes.

Since, according to Greimas, the city can be considered an object-message produced by a sender and read by a receiver, it constitutes a "text." The grammar of this text is to be constructed by the semiotician. Since the text is an agglomeration of people and things, the semiotician tries to observe recurring relationships between them that permit the construction of metatexts containing descriptions of the agglomeration of people and things in terms of inventories or sequences of "pronouncements." In these the users of the city are the grammatical subjects, and the things that they are in context with are the objects. Recurring observable relationships between subjects and objects together with their semantic interpretations constitute lists of typical recurring pronouncements.

The city space as a text is not only defined by its perceived sensual characteristics (visual, acoustic, thermal, olfactory, etc.), but also by the objects that support these qualities. Since these objects are products of human actions, the text is also defined by actions and meaningful behavioral programs. The latter may ultimately be substituted by automated programs. These aspects of sensation, objects of their support, and behavioral programs are illustrated by the following example: In order to get a positive thermal sensation, one has to either get wood to make a fire or switch on the central heating. The latter case of individual action depends in turn on a collective entity with its own object support, for example, the urban distribution of gas by a "collective subject"—the gas company. The individual case is constituted by the set of relations of the individual with objects, the individual being the center of these relationships. The collective case is made up of a set of networks (electricity, gas, water, sewage system, mail, subway, streets, etc.), the terminals of which are constituted by individuals.

The production and reading of the *object message city* as the set of relationships between subjects and objects involved should not be misunderstood as an activity by which urban designers send a message to citizen users, as designers sometimes would prefer. Rather, it is a complex process carried out by collective entities pronouncing and reading the city and which influence each other mutually.

SETTLEMENT SPACE

The Hjelmslev–Greimas model, with its subdivision of signs into a level of content and one of expression, has been used recently in a synopsis on meaning of settlement space (Lagopoulos, 1986). The author distinguishes between the semiotic production and the conception (consumption, perception) of settlement space and then differentiates the form of content and the form of expression of both perspectives. After describing the approach of Greimas, he briefly discusses the ideas of Agrest and Gandelsonas (1973; Agrest, 1977) in order to show what the subject matter of the semiotics of settlement space might be. Agrest and Gandelsonas hold that the object of semiotic analysis in architecture is a system for delimiting architectural configurations, which are, in turn, delimited by means of drawings, related texts, and the construction process itself. This system is also connected to other than architectural semiotic systems as Eco (1968) had suggested earlier. Since the boundaries of the system are open to other than architectural semiotic systems, they are also open to ideologies. Ideologies at the base of architecture are, therefore, a legitimate subject of semiotic analysis.

For Lagopoulos (1986), the constructed space of a settlement is part of the superstructure of a society, including its ideology. The base of the society determines space and ideology. The *production of space* consists not only of semiotic, but also of nonsemiotic processes (e.g., construction). The semiotic processes in the production of space are constituted by a set of connotative (symbolic) codes rooted in the ideology of the superstructure. There is a correlation between the modes of production and the dominant connotative codes stemming from the dominant classes of society as is the case, for example, when religion dominates society.

As far as the *form of the content level* is concerned, Lagopoulos observed that there are different ways of segmenting it. One way is to identify universal spatial characteristics, such as the topological, projective, and metric spaces that were proposed by Piaget and Inhelder (1948) in their theory of developmental psychology. Pinxten (1974, 1976, 1977) has suggested that there are 78 spatial terms that are relevant to the description of spatial meaning in any culture. These terms could, therefore, be considered universal semantic characteristics of space. They are identical with all topological and certain projective properties of Piaget's theory. This suggests that the form of content in spatial meaning is organized universally. This does not, however, exclude the possibility of spatial meaning also taking on additional culture-specific forms.

While the form of content consists of topological and projective properties of space for the *form of expression* in the production of space, geometrical form has been discussed as a possible basis of organization. Eco (1968) argues that geometry is a metalanguage and not the exclusive property of architectural expression, but Greimas (1974) and his school (Castex & Panerai, 1974; Castex & Depaule, & Panerai, 1978) hold, though differing in detail, that the form of spatial expression is geometrical. Greimas (1974) postulates a three-step hierarchy of spatial organization increasing in complexity. The first step consists of elementary units (e.g., straight or curved lines) that are put together according to the rules of an "elementary grammar." The second contains autonomous elementary structures (e.g., a square) composed by following the rules of a "phrase grammar." The third is made of configurations (of elementary structures) following a "text grammar." Elementary units, elementary structures, and configurations thus constitute a hierarchy of geometrical forms, according to which the level of expression of settlement space is organized.

Other authors (Boudon, 1973, 1977a, 1977b, 1978, 1981a, 1981b; Hillier & Leaman, 1973, 1975; Hillier, Leaman, Stansall, & Bedford, 1976) also believe in an autonomous logic of space. Boudon tries to formulate a grammar of places based on a deep structure of forms (like the deep structure of sentences in linguistics). Hillier and Leaman postulate a morphology of architectural and urban patterns based on elementary objects, their relations, and eight operations. These patterns are shaped in part by their own logic and in part of exogenous social variables.

After considering the production of settlement space, the second part of Lagopoulos' (1986) review is concerned with the *conception* (perception, reading, etc.) *of space.* While the first perspective is that of powers shaping space, the second is that of the user. The question to be answered is about the form of the content in "reading" space. Eco (1972) uses a method known in semantics as "componential analysis." He applies it to architectural objects (as linguists apply it to verbal material) to analyze the structure of architectural meaning. The architectural form of content is organized in "sememes," a term borrowed from semantic analysis of verbal meaning. For a "column," such elementary units of meaning could be "resting upon the ground" and "holding up." These sememes consist of a set of semantic markers (such as "solid material," "vertical extension," "Doric," etc.). Similarly, for Fauque (1973) the form of content of the urban image is organized in constellations of elementary semantic traits (which he calls "semes"). They are limited in number but capable of generating a large number of urban places. These places must be analyzed by semantic axes, which depend on culture and on specific groups. Therefore, according to Fauque (1973), the content of urban signs is anthropological.

The organization of the form of content is always described as correlated with a corresponding organization of the form of expression. For Eco (1972) the units of this organization are morphemes, again a term borrowed from linguistics. These are called "visual figurae" by Jencks (1980), in the case of a column, base, shaft, and capital. Fauque's (1973) organization on the level of expression is made up of what he calls *urbemes,* such as "squares" and "roads" with their markers grouped in oppositions (wide/narrow, dense/sparse). Polar changes in these oppositions correspond to changes in the meaning of an urban sign.

There is also a semiotic approach to the city with a more sociological perspective. According to Ledrut (1970, 1973), the city becomes meaningful through different visual and nonvisual semiotic systems and their connotative codes, which in turn form a global urban connotative system, a "world view" rooted in society. Because of the number of different codes involved that originate in various groups, no units of the form of expression can exist that are pertinent for all groups.

While the forementioned authors all define synchronic settlement structures, Choay (1970, 1970–1971) offers a diachronic perspective. She subdivides Western settlements into categories according to the amount of meaning they incorporate. Pre-Renaissance settlements were highly charged with meaning. After the Renaissance, settlement became increasingly "hyposignificant," (i.e., decreasing in meaning). Finally, due to industrial production, modern settlements are only "monosemic," (i.e. one-sided in meaning).

Urban Culture

In his article "Urban Culture," Gottdiener (1986) also applies the Hjelmslev–Greimas model. He follows the general line of analysis of settlement space proposed by Lagopoulos (1978, 1986), but renders it even more sophisticated. In modern society, space results from the articulation of three interacting structural systems: economics, politics, and ideology. This means that on both levels, content and expression, the signifieds and respectively the signifiers of several codes are often in conflict. Especially on the level of content, the analysis of the form of signifieds may become ambiguous because they belong to different codes at the same time. Likewise, the material realizations of a society in space might be multicoded and thus belong to several semantic systems (on the level of content). This is the problem that confronts the semiotics of urban culture.

Gottdiener finds three hints to the solution of this difficulty in the work of Barthes (1967).

First, urban space is a special case of the semiotics of objects where the units of analysis stand at a junction of separate codes. As a consequence, on the level of content, each form of the signified of space "is also the form of the signified in other ideological systems such as work, leisure, nationalism, class domination, and religion" (Gottdiener, 1986, pp. 1142–1143).

Second, code users know a variety of lexicons, corresponding to different social practices. As a result of this multicoding, different lexicons are applicable to the same object. Different sets of signifieds on the level of content might coexist in the same individual and determine a number of different readings.

Third, the metalanguage of interpretation applied by the analyst on the level of meaning could be socially and locally (and professionally!) specific and, therefore, give rise to different semiotic conceptions of the same space.

The problem of multicoding has been addressed in different ways. As mentioned above, Choay (1973) observed that the modern city is "hyposignificant," while, at the same time, it is multicoded on the level of the semiotics of consumption. Jencks (1980) differentiated between the elitist and technological code of architects and the more traditional code of the consumer, both engendering different interpretations of the same object. Agrest (1977) considered culture a hierarchy of codes in which the urban place is a node representing the intersection of various such codes.

ARCHITECTURAL SEMIOTICS

Preziosi's (1986, 1979a, 1979b, 1983a) work on the semiotics of architecture could be considered the most advanced in the study of environmental meaning from a semiotic point of view. He criticized the "linguistic fallacy" of early attempts looking for parallels between language and architecture. Architecture belongs to the semiotics of visual communication differing from language by the medium (air vs. virtually all material resources of an ecology usable for buildings), by dimensionality (temporally vs. spatio-temporally), by permanence of the stimulus (fading vs. permanence of broadcast), and instrumentality (voice vs. whole body and instruments). Equally, Preziosi is in agreement with Eco (1968) on the "artistic fallacy" considering only works of art (or artists) the subject matter of architectural semiotics. He equally criticizes the limited range of functions considered in many theories based on Hjelmslev's restrictive concept of denotative and connotative signs, advocating a multifunctional approach to architecture as had been proposed earlier by Mukarovsky (1937–1938) and later by Jakobson (1960). Mukarovsky had postulated that a building could have functional, historical, social, aesthetic, and individual functional meanings. Jakobson called his six functions of linguistic messages the emotive, referential, conative, aesthetic, phatic, and metalinguistic. Preziosi (1979a, 1979b) adapted these six functions to architecture. He called Jakobson's emotive the expressive, his referential equally the referential, his conative the exhortative, his aesthetic equally the aesthetic, his phatic the territorial, and his metalinguistic the allusory functions of architecture.

Preziosi also criticized the "sender–receiver fallacy," closely related to the "artistic fallacy" making the architect into an active sender and the user into the passive receiver. Related to this is the "building fallacy," that is, considering the study of such traditional categories as buildings or urban objects. The study of architectural semiotics coterminous with the study of architecture proper is for Preziosi one of history and theory of institutions, whereas the study of semiotics of architecture addresses artifacts in direct and indirect relation to bodily instrumentality and the

actively seeing subject. In the framework of architectural multifunc-
tionality, form emerges as a site of meaning production by subjects.
Meaning production is based on the perception of those formal equiv-
alences and disjunctions that mark equalities or differences in meaning.
In fact, semiosis and perception are two sides of the same coin. In
acculturated individuals, perception is a semiotic activity. The built environment is not perceived as a collection of stimuli but as meaningful
patterns of forms, i.e., as a constellation of signs. The environmental
array comprises a hierarchical network of signs formed by a cascade of
relational interactions of edges and contrasts of shapes, colors, textures,
materials, sizes, and scales. This equivalence of semiosis and perception
in Preziosi's architectural semiotics opens it up to empirical verification
by methods used in perceptual and cognitive psychology.

On the basis of systematic examinations applied to large samples of
mostly Minoan architecture, Preziosi (1979b, 1983a, 1983b) proposed a
system of architectonic signs. This architectonic code comprises two
levels: sense-discriminative signs marking only differences, and sense-
determinative signs having a meaning of their own. Three types of
sense-discriminative signs are distinctive features, forms, and templates. Distinctive features, are, for example, the relative height, length,
or width of an architectonic object, such as a wall. Forms are made up of
simultaneous clusters of distinctive features. Templates, in turn, consist
of sequential arrays of forms. Buildings are perceptually palpable
through these juxtapositions in sequence. So far, about 20 forms have
been found that can be distinguished by contrastive features such as
ratios of height, length, width, and oppositions of color, texture, and
material. Some of these visually distinctive features may also have a
sense-determinative function, such as color oppositions in flags, not
only to mark a "difference," but also to refer to a given city or country.

As far as sense-determinative signs in the architectural code are
concerned, Preziosi distinguished four types, which he called figures,
cells, matrices, and settlements. Figures are made up of one or more
sense-discriminative signs (i.e., distinctive features, forms, or templates) composed simultaneously on a surface or sequentially in a row.
Cells consist of one or more figures. Matrices are patterns of aggregation
of sense-determinative units comprising one or more cells. Settlements,
finally, contain one or more matrices.

It is important to understand that these hierarchized differences in
signifiers are not necessarily correlated with equally hierarchized architectural signifieds, such as rooms, suites, buildings, and so on. The
architectural code is also not based on size but on the modes of sign
production and sign use. Thus, a settlement may be coterminous with a

cell, matrix, or figure. These levels of sign types represent various ways of perceiving semiotic formations composed of clusters or relationships.

Since these formations are addressed to visual perception and discrimination, it should not be a surprise if they are hierarchized similarly to the signifiers discriminated by other sense modalities. Thus, in language addressed to acoustical discrimination, we have similarly sense-discriminative and sense-determinative units. It is not a relapse into the "linguistic fallacy" if Preziosi establishes parallels between the distinctive features of the architectural and the linguistic codes, between form and phonemes, between templates and the alternation of vowels and consonants in syllables, between figures and morphemes, cells and words, matrices and phrases, settlements and texts. As Rossi-Landi (1968, 1972, 1975) has shown for objects in general, this parallelism stems from the common characteristics of production in generating human artifacts.

Conclusions on the Semiotic Approach

Looking over the whole range of concepts in the field of spatial and architectural meaning, it can be said that these concepts have become more and more sophisticated during the last decade. One reason seems to be their increasing emancipation from the models of the different schools of semiotics. Another reason is the discovery of multicodal and multifunctional complexity of spatial meaning. The realization that the process of sign interpretation is the same as that of form discrimination in perception and cognition bridges the gap originally dividing the approaches of semiotics and the social sciences to the problem of environmental meaning.

THE ORIGINS OF ENVIRONMENTAL PSYCHOLOGY AND ITS RESEARCH ON ENVIRONMENTAL MEANING

Environmental psychology and its subset, architectural psychology, which both investigate environment–behavior relations, can be traced to Lewin's work on life space and Barker's investigations of behavior settings. Research on environmental meaning in this field was begun in the 1960s, when in anglophone countries, especially in the USA, the interest in design methods was complemented by the search for their meaningful application. Architects, sociologists, and geographers in such interdisciplinary organizations as the Environmental Design Research Association (EDRA) in the USA and the International Association for the

Study of People and Their Physical Surroundings (IAPS) in Europe soon began to undertake research on environmental meaning. The proceedings of the conferences of these organizations regularly published studies on the topic, as did the two periodicals *Environment and Behavior* and the *Journal of Environmental Psychology.* Some of the pioneering work of Appleyard, Canter and Tagg, Harrison, and Howard, Hershberger, Honikman, Krampen, Kreimer, Lowenthal and Riel, Muntanola Thornberg, Sanchez-Robles, and Stringer was reprinted or published for the first time in *Meaning and Behavior in the Built Environment* (Broadbent, Bunt, & Jencks, 1980). This volume closed with "A Semiotic Program for Architectural Psychology" by Broadbent.

Studies on spatial meaning in environmental psychology are not as easy to find, because that is not a main thrust of research in the field. This is evident in the early reader on *Environmental Psychology* edited by Proshansky, Ittelson, and Rivlin (1970), which contains only two articles in which environmental meaning is explicitly treated (Beck, 1970; Ruesch & Kees, 1970). Wohlwill's (1977) overview on environmental psychology in the *International Encyclopedia of Psychiatry, Psychology, Psychoanalysis, and Neurology* mentions environmental meaning only in passing under the heading "Recurrent Issues: Influence of Physical Environment on Behavior": "concepts such as mental maps, personal space, environmental meanings and values, and theoretical frameworks such as those of cognitive-developmental and personal-construct theory have been invoked to handle phenomenon in this field" (Vol. 4, p. 340). Even in the series *Human Behavior and Environment* started by Altman and Wohlwill, the problem of environmental meaning has only been raised tangentially in recent volumes (Duncan, 1985; Hunter, 1987; Lawrence, 1985). Neither is there a chapter on environmental meaning or symbolism in the very recent *Handbook of Environmental Psychology*, edited by Stokols and Altman (1987). Nevertheless, in the proceedings of EDRA and IAPS conferences, the topic of environmental meaning has been treated consistently over the years. There are also two monographs in which the theme has been explicitly treated, one from a psychological point of view (Krampen, 1979a), and the other from an anthropological point of view (Rapoport, 1982).

ENVIRONMENTAL MEANING IN ENVIRONMENTAL PSYCHOLOGY

According to Miller (1967, 1969), an indirect approach to (verbal) meaning would utilize the assessment of "semantic distances" between words. This hypothesis can be generalized also to environmental meaning: for instance, places with a small semantic distance are similar in meaning. Semantic distances can be estimated by various measuring devices. The first approaches in this direction were carried out applying the semantic differential of Osgood, Sucy, and Tannenbaum (1957) to environmental objects by measuring their similarity on the three semantic dimensions of evaluation, activity, and potency (Berger & Good, 1963; Krampen, 1971, 1979a; Sommer, 1965).

The use of the standardized version of the semantic differential was soon criticized (Bechtel, 1980) and replaced by a search for the proper dimensions of architectural and spatial experience (Hershberger, 1973; Kasmar, 1970) in order to construct scales with items adequate to their objects. These were applied to facades (Bortz, 1972; Krampen, 1974, 1979a), entire sections of cities (Franke, 1969), and landscapes (Bauer, 1980; Bauer & Bräunling, 1982; Wohlwill, 1976; Zube, 1976).

The question arose whether facing the subjects with ready-made scales of a semantic distance was the best approach to tap their environmental values and cognitions and to assess environmental meaning. This led to the application of the Repertory Grid method (Kelly, 1955) to get at the personal constructs people had about environmental objects and situations (Honikman, 1973; Sanchez-Robles, 1980: Stringer, 1980).

Since generalizing from personal constructs entails difficulties, other more sophisticated methods were sought and found. One of them was the free sorting of environmental objects (or their pictures and a subsequent cluster analysis of the subjects' sortings; Krampen, 1979a).

One of the most sophisticated techniques applied to the study of environmental meaning is the multiple-scaling method, with a subsequent treatment of the data by smallest-space analysis (Lingoes, 1973). This technique has, for example, been applied by Canter and Tagg (1980) to the designative, appraisive, and prescriptive dimensions of meaning (Morris, 1964) of various building aspects and their attributes. Krampen (1979a) applied smallest-space analysis to cluster distinctive features of buildings (e.g., relative size, complexity, number, and size of windows) around building types (office, factory, church). Groat and Canter (1979) and Groat (1982) used it in the grouping of building styles. All these techniques have advantages and disadvantages, but the question of what empirical research means without the guidance of theory remains.

ENVIRONMENTAL COGNITION

There is at present no large-scale social science theory of environmental meaning available. The questions of environmental meaning are sporadically treated, however, within the context of cognitive theory under the heading of environmental cognition (Moore, 1979; Moore & Golledge, 1976). The basic assumption of the microtheory of environmental cognition is that "environment" is a mental construct, i.e., a cognitive representation. The content of cognitive representations is variously described in terms of Lynch's (1960) paths, edges, nodes, and landmarks, or in sociological categories—the city as a center of civilization, of economic opportunity, as a melting pot, as a place for superficial contact or of depravity, and so on. People's cognitions can be described in terms of the relationship between "objective" descriptions of the environment and subjective knowledge (e.g., cognitive mapping), subjective constructs, images, and, finally, in terms of the meaning the environment has for them. Thus, part of the cognitive construction of the environment is giving meaning to it. And, therefore, meaning is only a partial aspect of environmental cognition. In fact, Moore (1979) bemoans the fact that this aspect is understudied. The logical consequence of the environmental cognition approach is an emphasis on individual and group differences. This applies also to meaning attribution (if it is studied at all). These differences are investigated in terms of content aspects and structural organization, between individuals and within individuals, in terms of cognitive style and stages of ontogenetic development, related to such explanatory variables as cognitive ability, age, sex, length of permanence in a place, walking or traveling, ethnic and cultural groups, and the like.

Moore (1979) summarizes the bulk of research on individual and group differences as follows:

> It seems to me that any attempt to understand the environmental psychology of urban life and the role of environmental cognition in experience and behavior must have as one of its goals the understanding of how people conceive of their environment, and the meanings which the city has for different people. People differ in their conceptions of space not only in quantitative and concrete terms but also in qualitative and social-symbolic ways. One person's "slum" is another person's "urban village," and one person's "ticky-tacky little boxes" is another person's dream castle and the badge of entry into the middle class. (p. 57)

According to Moore (1979), there has been a lack of studies on environmental differences, where not individuals, but physical settings are assessed on those differences that prompt cognitive representations, meanings, and memories. Appleyard (1969) isolated three such major environmental factors: intensity of certain uses, visibility by commanding location, and uniqueness of physical form.

The Organismic-Developmental Perspective

As in environmental cognition, environmental meaning plays a role that coincides with other factors in other theoretical frameworks. The organismic-developmental theory (Wapner, Kaplan, & Cohen, 1980), for instance, proposes to understand the relationship between people and environment as a purposeful transaction. It focuses on three themes: first, the teleological character of all action; second, the contextual and holistic determination of any specific action; and third, the genetic-structural ordering of the modes of action in various living beings under various conditions. The instrumentalities related to goal-directed action taking place in a social, cultural, and physical context are perceiving, thinking, and symbolizing. These processes are interdependent and interfunctionally related. The construal of the environment by means of these processes depends on the agent and varies with his or her attitudes and states. Thus, the environment is perceived, cognized, and symbolized depending on the agent and his or her goals. Environmental symbolism, i.e., meaning, therefore, depends on goal-directed action.

A Microtheory of Environmental Meaning

In my earlier work (Krampen, 1979a), I proposed a theoretical model of environmental production and consumption based on cognitive conceptualization and perceptual recognition (i.e., reconceptualization) of environments. The process of conceptualization goes through the cognitive establishment of two semiotic structures, one containing functional (denotative), the other stylistic (connotative), meaning. A semiotic structure is a meaningful system in that it consists of two sets of coordinated classes, that is, two classification systems. One classification system contains the classes of meaning carriers or signifiers. The other consists of classes of meaning or signifieds, each one of them coordinated with one class of meaning carriers. Since the classification system of signifiers is coordinated with a classification system of signifieds, and each class of signifier is coordinated with one class of signified, the resulting meaningful structure may also be called a *sign system*, a sign being a unit of a signifier with its signified (Figure 5).

Now consider first the process involved in planning or producing a meaningful environment. In any conceptualization, hence also in the conceptualization of a meaningful object, this act of double classification (in classes of signifiers and signifieds) must be gone through twice (Figure 6). At first, the denotative meaning of an object is established; in other words, the choice of *what* the object is must be made by selecting classes of signifiers   classes of signifieds

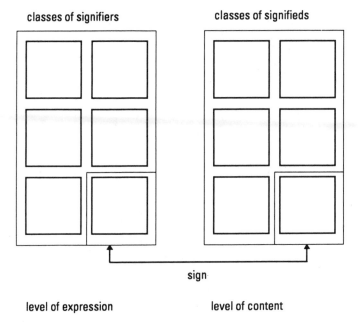

sign

level of expression   level of content

FIGURE 5. A semiotic structure consisting of a classification system of signifiers and a classification system of signifieds. A sign is the coordination of a class of signifiers with a class of signifieds.

one of the signifier classes and its coordinated class of signifieds. For example, out of the semiotic structure holding the classes of all social building functions, say offices, factories, churches, schools, high-rise apartments, family houses, and so on, one functional class is chosen, say, the class of all buildings having the function of a family house.

It is obvious that the conceptualization process is not complete at this point. In a second round of double classification the connotative meaning has to be established, that is, *how* the chosen function is implemented must be selected. This selection is a question of "stylistic" meaning, referring to the term "style" in its broadest sense. For example, out of the semiotic structure of all the different manners in which a family house can be realized, say, wood, steel, and glass, concrete, brick, functionalist, traditional, and so on, one "style" is chosen, say wooden traditional.

This twofold double classification process in the conceptualization of a meaningful object cannot be reversed. One cannot, obviously, an-

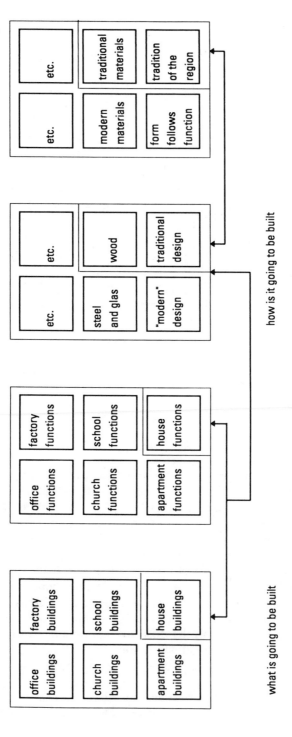

FIGURE 6. The twofold double classification in the conception (and perception) of buildings. At first it is decided *what* the function of the building is going to be. Then it is decided *how* this function is going to be implemented.

swer the question of how anything is going to be realized before it is clear what its function is. But from the moment in which the question of what it is going to be has been answered, the question of how it is going to be implemented cannot be escaped. In the conceptualization process the term "denotative function" does not designate a monofunctionalist concept, as it may seem from the above example. A building has to satisfy quite a number of functions; for example, in its social function as a house (and not a factory), it has to be statically sound and it must have enough rooms to satisfy the territorial needs of a family. Likewise, the term "stylistic function" does not refer to a narrow concept such as art historians may have. Both concepts are meant to be multifunctionally open.

So far the conceptualization process has been treated from the perspective of planning or producing a meaningful environment. The same process, however, can be looked at, secondly, from the perspective of the observer or user. In this case, what has been conceptualized by the producer is reconceptualized by the user. The user is not a passive consumer of what the producer is offering, but actively searches his or her way through the environment. If looking for an office building the observer has, for the moment, ruled out all other social functions in his or her system of denotative functions of the environment, and only looks now for denotative and stylistic features of office buildings. The same is true in the commissioning of a home to an architect (not a factory or a church). Thus, the user of meaningful objects goes through the same process as the producer, first conceptualizing denotative and then connotative functions.

The semiotic structures of denotative and connotative signs as described for buildings present only a synchronic perspective. From a diachronic perspective, their changing over time must be studied. This could be done, for example, from a historical-dialectic point of view.

Different denotative and connotative functions of buildings are conceptualized and reconceptualized by means of distinctive features, that is, of characteristics serving to distinguish one building class from another. If this was not the case, all meaningful objects would look alike. In one study, respondents were shown four series of building pictures, one set was photographs from which the other three series had been abstracted by drawing: the first by rendering only the external silhouette, the second by adding the storeys, and the third by adding the windows (Krampen, 1989). In four separate sessions spaced two weeks apart, respondents saw the first series with minimal detail, then the second and third with more detail, and finally the photographs.

They were asked to note the social function of the building and the features by which they recognized it, and to rate how secure they felt about their judgment. The result was that the number of different building functions named for each picture across respondents decreased and the degree of security increased significantly as detail in the pictures was increased from session to session. When detail was very scarce, the information used was building size and height, the form of the roof, and additional features such as presumed steeples or smokestacks. But with the addition of stories in the buildings pictured, windows were mentioned more frequently, the story line probably being taken for part of (large) windows. Windows were mentioned most frequently at the third stage when they were introduced in the drawings. In the photographic pictures, such details as curtains at the windows or symbols (crosses, writings) were used to discriminate building functions. The result of this study seem to indicate that the social functions of buildings are recognized by combining mutually exclusive sets of distinctive features of buildings as signifiers with mutually exclusive sets of distinctive features of social functions in the coordinated classification system of the signifieds.

This study led to the design of another (Krampen, 1989), in which respondents were asked to estimate the probability of encountering a given set of distinctive features when imagining buildings with different social functions. The functions to be imagined were office, factory, church, school, apartment building, and family house. For the contours of the buildings, the probability of occurrence of the following four distinctive features was to be estimated: building size (large vs. small), secondary buildings such as steeples, smokestacks (present vs. absent), direction of building shape (horizontal vs. vertical), and form of roof (flat vs. gabled). For the internal structure of the facade, the following four distinctive features were given: storeys (many vs. few), number of windows (many vs. few), window size (large vs. small), and direction of window shape (horizontal vs. vertical). For each of the four distinctive features of the contour and the facade, the probability ratings of the respondents were averaged ($N = 30$). In each column of percentages over the six building functions, no equal profiles were found. This suggests that all six building types had been distinguished by the eight features. By transforming the percentage values into mutually proportional measures of size (e.g., facade surface and height) and frequency (e.g., number of windows), phantom drawings of the six building types could be prepared, which gave an impression of their average image (Figure 7).

The conclusion of the two studies is that respondents identified the

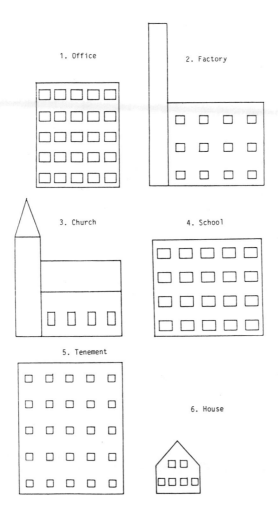

FIGURE 7. Phantom drawings resulting from estimated probabilities of a certain building's size, verticality, having a secondary addition, an elaborated roof, many stories, many windows, large windows, and vertical windows.

different social functions of buildings with a fairly high degree of security if sufficient distinctive features were presented. Also, respondents seemed to have internalized stereotypes of building types and bundles of distinctive features which served as "templates" for the recognition of building functions.

Conclusions on the Environmental Psychology Approach

Considering the development of meaning studies in environmental psychology, a quantitative increase in the number of investigations and a qualitative improvement must be noted. Methods have become increasingly sophisticated as the problems have increased in scope and variables. The main thrust of research has produced knowledge about individual differences in environmental cognition and transaction. There is an increasing awareness in environmental psychology that environment–behavior relationships have to be treated holistically and are never devoid of a meaning component. Moreover, not only individuals must be assessed on how they construe their environment cognitively, but also environments must be assessed as to the properties that mediate certain meanings for different people.

OUTLOOK ON THE FUTURE: ENVIRONMENTAL MEANING IN AN ECOLOGICAL PERSPECTIVE

During the last decade, a new perspective in the study of environmental meaning has emerged. It is not a closed and finished view of the problem but a sketch of a possible solution that requires further elaboration. This new perspective is presented in Gibson's (1979) ecological approach to perception. It seems that an ecological approach will reflect the reciprocal relationship of the organism and its environment more adequately than either treatment of environment as a category of perception or cognition or as a set of objective data. If this approach holds, the subject–object split, plaguing both the semiotic and the psychological perspectives on environmental meaning, eventually would be superseded. To begin with, I shall clarify some fundamental concepts of the ecological approach.

The first and most important concept to be defined is *ecology*. It should be used in the sense that biologist Haeckel gave it when he first introduced the term in his 1866 work "General Morphology of Organisms": "Ecology is the general science which studies the relationships of the organism and its external environment." Would an ecological approach to the study of environment be the investigation of the organism's relationships with its surrounding "space"? But then, what does "space" mean in an ecological perspective? Certainly not the ghostlike line grid of the three-dimensional coordinate system; that is intellectual space. Ecological "space" is lived-in and moved-through by the organism. Ecologically speaking, the organism is continuously moving through a penetrable medium without hitting the surrounding surfaces.

Further, there are important differences between the concept of *scale* in the world as described by physics and that of the environment of animals and humans. The physical world ranges from galaxies to atoms. Our lived-in, visible environment is composed of all we must see in order to act successfully. As a matter of fact, we see only what we must see: landscapes, mountains, trees, plants, buildings, objects, animals, and, last but not least, people. Our environment ranges from textures measurable in millimeters to objects measurable in meters to the visible landscape measurable in kilometers. We cannot see galaxies or atoms.

Organism and environment form an inseparable whole. No organism can exist without its environment and vice-versa. Hence, what was present before life began on earth cannot be called environment. A good simile for this *organism–environment mutuality* is the well-known Chinese duality of yin and yang.

Our terrestrial habitat is best described by its *medium*, its *substances*, and its *surfaces*, which separate the substances from the medium. The gaseous atmosphere on earth is the medium of many animals and of humans. It permits locomotion, seeing, smelling, and hearing. Substances are matter in a solid or semisolid state. They do not permit movement. They differ in hardness, viscosity, density, elasticity, plasticity, and the like. All substances have surfaces that are layed out in the environment. Surfaces and their layouts tend to resist deformation. Depending on the composition of the substance they delimit, surfaces have different textures and pigments. Surfaces have shape, may be more or less lit, and may absorb or reflect more or less of the illumination falling on them.

AN ECOLOGICAL APPROACH TO MEANING: THE THEORY OF AFFORDANCES

The environment of the organism is meaningful with respect to the organism and its scale. For example, a rock 45 centimeters high affords seating to the grownup wayfarer; to the accompanying child, the rock affords a table. In relation to the body size of grownups and children, the affordances of seating and table top remain a constant meaning of that rock.

The concept of *affordance,* coined by Gibson (1979), is the ecological equivalent of meaning. The idea that the meaning of a thing has a physiognomic quality (as have the emotions that appear in a person's face) stems from Gestalt psychology. Koffka wrote in his *Principles of Gestalt Psychology:* "Each thing says what it is . . . a fruit says 'Eat me'; water says 'Drink me'; thunder says 'Fear me' (1935, p. 7). The handle "wants to be grasped," objects 'tell us what to do with them" (p. 353). Koffka called this the "demand character" of things.

Lewin, considered by many as the founder of ecological psychology, used the term *Aufforderungscharakter* (Lewin, 1926), translated into English as "invitation character" or "valence" to express the meaning of things.

Von Uexküll (1913), the precursor of ethology, described what he called the "counterability" (*Gegenleistung*) of human-made things, in this passage on a stroll through town:

> It is not without interest to start a stroll through the town if one remains conscious of a certain question while looking at things. Thus, we want to ask which meaning have objects striking the eye and for whom do they have a meaning? . . . Everything—indeed everything which we get to see is adapted to our human needs. The height of houses, of doors and windows can be reduced to the size of the human figure. The stair fits our gait and the bannisters the height of our arms. Each single object is endowed with sense and form by some function of human life. We find all over an ability of man which the object sustains by its counter-ability. The chair serves seating, the stair climbing, the vehicle riding, etc. We can talk about something being a chair, a stair, a vehicle without misunderstanding, because it is the counter-ability of the human products which we really mean by the word which denotes the object. It is not the form of the chair, the vehicle, the house which is denoted by the word, but its counter-ability.

There is optical information available to the organism for perceiving affordances. In fact, perceiving has never to do with value-free objects to which meaning is later associated. All ecological objects are full of meaning to begin with. Surfaces, their layouts, and the substances they delimit always show affordances for someone. Affordances may be positive or negative. A stair invites stepping down by its small height, a cliff warns not to step down because of the abyss at its rim. The information specifying the positive or negative affordances is always accompanied by information specifying the perceiving organism itself, its body, legs, hands, nose, and so on. In fact, we cannot perceive the environment without perceiving ourselves within it. This shows once more the yin-and-yang nature of the organism–environment relationship.

In special cases there may be misinformation issuing from certain objects, such as large glass windows, which to birds may appear to be a medium to fly through. And many of us have in a hurry bumped our heads against a glass door, mistaking it for an opening.

Some Affordances Which Are Independent of Surface Layout

According to Gibson, there are some affordances in the environment that depend less than others on information issuing from surface layouts. For instance, water, a necessity for terrestrial organisms, is not recognized by a steady surface. Since body tissues consist to a large extent of water, we must constantly replenish the liquid by drinking. The organism must, therefore, recognize water. Fire, instead, is an event having a beginning and an end, affording warmth and illumination, allowing one to cook meals, melt metals, and the like, while consuming fuel substance. But it also has negative affordances if one gets too close to it. Both water and fire have many kinds of uses, and therefore, many meanings to humans.

Animate objects differ from inanimate ones in that they actively move by means of their own internal forces. They may be prey or predator, mate or rival, old or young. They may afford eating or being eaten, care or aggression. What one animal affords the other is also social interaction. Social interaction requires coordination and cooperation. This part of Gibson's theory of affordances is still to be developed.

The Meaning of Surfaces and Their Layouts

A nomenclature for surface layouts. Most affordances are directly specified for the organism by surfaces and their layouts. Gibson (1979, 1982) provides a systematic nomenclature of different kinds of layouts. One is the *ground*, referring to the terrestrial surface, and implying the effect of gravity, a horizon, and the sky. If there was only the ground, the layout would specify an *open environment*. But this condition is only realized in a flat desert. Generally, the environment is full of convexities and concavities and all kinds of "clutter." An *enclosure* is a layout of surfaces that surround the medium. The totally enclosed medium is seldom realized, as, for example, in the case of an embryo. A *detached object* consists in a layout of surfaces that is entirely surrounded by the medium. Examples of detached objects are all moving animals, including humans, but also balloons. Many objects seem to be *attached*, that is, practically surrounded by the medium. Most of them are attached to the ground. A *partial enclosure* such as a concavity, a hole, or a cave consists of a surface layout that partly encloses the medium. A *hollow object* is an object from the outside but an enclosure from the inside. Examples are a snail's shell, a hut, or a pitcher. A *sheet* consists of two surfaces that enclose a substance but are very close together in relation to their dimensions. A *fissure* is a layout of two parallel surfaces that are very close together in relation to their size and enclose the medium in a thin opening, such as a crack in a boulder.

A *place* is defined by Gibson as a location in the environment, a more or less extended surface or layout. Places have a name, but no sharp boundaries. They may be located by inclusion into larger places (e.g., the fire place in the living room of the Villa Camaioli in Florence). The *habitat* of animals and humans consists of places.

There are some more terms given by Gibson for surface layouts, but for the purpose at hand the above will suffice. These types of surface layouts are the signifiers of the environment—what they afford is the signified.

What surface layouts afford.   The affordances of *terrain features* either facilitate or prevent locomotion for animals and humans. Paths afford the locomotion of pedestrians from one place to the other. Obstacles, barriers, water margins, and brinks are negative affordances with respect to locomotion. Slopes may or may not permit locomotion depending on their angle. For thousands of years humans have been changing the terrain by constructing roads, stairways, and bridges to facilitate locomotion. They have also constructed walls, fences, and other obstacles to prevent access to their enclosures.

Different *places* in the habitat of animals or humans have different positive and negative affordances. Some places serve as refuges from predators, others as homes, which are partial enclosures. There are places to hide (hiding means to position the body in such a way that observers may not see it). Privacy in the design of housing means to provide opaque enclosures in order to prevent others from looking in. Curtains on windows permit or restrict light and permit seeing without being seen.

Since the atmospheric medium changes from warm to cold and from rain to snow, humans must have *shelters*. Originally they used caves, but then they started constructing artificial shelters, so-called huts. Huts are objects attached to the ground. They feature a roof that affords protection from rain, snow, and sunlight, walls that shelter against the wind, and a doorway that permits entry and exit.

Objects are of persisting substance with closed or nearly closed surfaces. They can be attached or detached. An attached object may be carried, and if it is of an appropriate weight it affords throwing. Hollow objects can be used as containers. *Tools* of different affordances can be considered extensions of human limbs, especially of the hand. Humans have developed *display surfaces*, such as pictures, which afford visual information. This information is secondhand, since it permits the beholder to see as a surrogate what the maker of the picture saw in the original.

A whole set of affordances typical for a given species may be called its ecological niche. Whereas the term *habitat* refers to the set of places where a species lives, the term *niche* means how it lives.

CONCLUSIONS ON THE ECOLOGICAL APPROACH

The ecological approach to the question of environmental meaning seems to render more concrete and even existential the question students of semiotics and environmental psychology have been asking for many years about the meaning of architecture and the urban environment (Krampen, 1979a). We no longer look at architecture and the urban environment "objectively," as if we were separated from them. We are forced by the ecological perspective to consider ourselves as part of our human environmental niche. Semiotics has insisted too much on the role of signs in communication, that is, in the transmission of secondhand experience. The firsthand experience of ecological meaning of affordances in the environment is, perhaps, a more difficult subject of study before us, but also a more fascinating one. This holds especially for the direct experience of aesthetic meaning, about which there still remains much to be discovered. Environmental psychology, on the other hand, has accentuated individual differences in environmental cognition at the expense of discovering common traits in the ecological niche of the human species. Perhaps the inescapable presence of the perceiver in perception could teach us that we are not subjectively removed from, but part of, the environment in which we see the meaning of things directly.

The foregoing discussion of the semiotic and psychological approaches to environmental meaning and their confrontation with an ecological approach implies some drastic changes for future research.

IMPLICATIONS FOR FUTURE RESEARCH AND APPLICATIONS

One result emerging from the foregoing review is the inadequacy of the denotative–connotative meaning dichotomy. Environmental meaning seems to be intrinsically multiple, as are environmental affordances. This multiplicity of affordances results directly from the organism–environment reciprocity, as was shown in the example of the rock serving as a seat to the grownup, and as a table to the child. This implies that much of the ergonomic literature could be unearthed and reinterpreted in the light of the theory of affordances. One example of the rehabilitation or ergonomics under the ecological approach is the study on visual guidance of stair climbing by Warren (1984). But not only ergonomic measurement should be taken into account. There are probably strong age and gender differences in the organism–environment reciprocity (Warren, 1984).

Since the term affordance refers to physiognomic perceptual properties of (designed) objects, there is no reason to confine it to lower-level meanings. Meanings on an intermediate or high level should also incorporate physiognomic properties, even if they are of the cultural or aesthetic variety. As a matter of fact, a theory of aesthetic affordance is needed. It might be constructed on the fact that the perceiver is always present in perception and that a comparison between body scale and the scale of the aesthetic scene may give rise to the collative variables (Berlyne, 1960) of surprise, contradiction, and other aesthetic effects (as is the case upon entering a Gothic cathedral). A whole new impulse would be given to the study of environmental aesthetics and meaning if this was conceived of as organism–environment fit.

Further research is needed on the correlation of perceptual architectural form codes, such as proposed by Preziosi (1979b, 1983a, 1986), and affordances. Would it be possible to reformulate Preziosi's architectural signs (figures, cells, matrices, and settlements) in terms of surface layouts as signifiers, and their multiple affordances as signifieds? Which forms are perceived as promising which affordances?

Obviously, the ecological approach to meaning requires changes in methodology. Field studies will have to replace, to a large extent, laboratory experimentation. How do perceivers move through and act in natural or human-made environments? This implies, in turn, a change in design education from the fixation on graphic interpretation of objects to the simulation of the visual experience of users moving from scene to scene in imagined future environments.

All this may find application in a way of designing environments. The goal of design should no longer be objects or "spaces," but programs for wayfinding, orientation, and visually guided action.

REFERENCES

Agrest, D. (1977). Design versus non-design. *Communications* (Sémiotique de l'espace), *27*, 79–102.

Agrest, D., & Gandelsonas, M. (1973). Critical remarks on semiology and architecture. *Semiotica, 9*(3), 252–271.

Appleyard, D. (1969). Why buildings are known: A predictive tool for architects and planners. *Environment and Behavior, 1,* 131–156.

Barthes, R. (1964). Eléments de sémiologie. *Communications, 4,* 91–135.

Barthes, R. (1967). *Elements of semiology* (C. Smith and A. Lavers, Trans.). London: Cape.

Bauer, F. (1980). Zur Konzeptspezifität des Semantischen Differentials.—Eine Diskussionsbemerkung zu Flade's: Die Beurteilung unweltpsychologischer Konzepte mit einem konzeptspezifischen und einem universellen Differential. *Zeitschrift für experimentelle und angewandte Psychologie, 27,* 163–167.

Bauer, F., & Bräunling, H. (1982). Ein Vergleich der Eignung konzeptspezifischer und universeller Formen des Semantischen Differentials zur Beurteilung von Umweltausschnitten. *Zeitschrift für experimentelle und angewandte Psychologie, 29*(2), 181–203.

Bechtel, R. B. (1980). Architectural space and semantic space: Should the twain try to meet. In G. Broadbent, R. Bunt, & T. Llorens (Eds.), *Meaning and behavior in the built environment* (pp. 215–222). New York: Wiley.

Beck, R. (1970). Spatial meaning and the properties of the environment. In H. M. Proshansky, W. H. Ittelson, & L. G. Rivlin (Eds.), *Environmental psychology: Man and his physical setting* (pp. 134–141). New York: Holt, Rinehart & Winston.

Berger, A., & Good, L. (1963). Architectural psychology in a psychiatric hospital. *Journal of the American Institute of Architecture,* December, 76–80.

Berlyne, D. E. (1960). *Conflict, arousal, and curiosity.* New York: McGraw-Hill.

Blomeyer, G. R., & Helmholtz, R. M. (1976). Semiotics in architecture. *Semiosis, 1,* 42–51.

Bortz, J. (1972). Beiträge zur Anwendung der Psychologie auf den Städtebau. II. Erkundungsexperiment zur Beziehung zwischen Fassadengestaltung und ihrer Wirkung auf den Betrachter. *Zeitschrift für experimentelle und angewandte Psychologie, 19,* 226–281.

Boudon, P. (1973). Recherches sémiotiques sur le lieu. *Semiotica, 7*(3), 189–225.

Boudon, P. (1974). Définition (sémiotique) des critères pour une théorie des lieux: Architecture et méthodologies. *MMI Bulletin des Séminaires Pédagogiques* (Paris: Institut de l'Environnement), 4, 3–57.

Boudon, P. (1977a). Introduction. *Communications* (Sémiotique de l'espace), 27, 1–12.

Boudon, P. (1977b). Un modèle de la cité grecque. *Communications* (Sémiotique de l'espace), 27, 122–167.

Boudon, P. (1978). Réécriture d'une ville: La Médina de Tunis. *Semiotica, 22*(1/2), 1–74.

Boudon, P. (1981a). *Introduction à une sémiotique des lieux. Ecriture, graphisme, architecture.* Montréal: Les presses de l'université de Montréal. Paris: Editions Klincksieck.

Boudon, P. (1981b). Recherches sémiotiques sur la notion de 'lieu architectural,' Review Article. *Recherches Sémiotiques/Semiotic Inquiry, 1*(4), 393–413.

Broadbent, G. (1969). Meaning into architecture. In C. Jencks & G. Baird (Eds.), *Meaning in architecture* (pp. 50–75). London: Barrie & Rockliff.

Broadbent, G. (1975). Function and symbolism in architecture. In B. Honikman (Ed.), *Responding to social change.* Stroudsburg, PA: Dowden, Hutchinson & Ross.

Broadbent, G. (1980a). The deep structures of architecture. In G. Broadbent, R. Bunt, & C. Jencks (Eds.), *Signs, symbols, and architecture* (pp. 119–168). New York: Wiley.

Broadbent, G. (1980b). Building design as an iconic sign system. In G. Broadbent, R. Bunt, & C. Jencks (Eds.), *Signs, symbols, and architecture* (pp. 311–331). New York: Wiley.

Broadbent, G. (1980c). A semiotic program for architectural psychology. In G. Broadbent, R. Bunt, & T. Llorens (Eds.), *Meaning and behavior in the built environment* (pp. 313–359). New York: Wiley.

Broadbent, G. (1980d). General introduction. In G. Broadbent, R. Bunt, & C. Jencks (Eds.), *Signs, symbols, and architecture* (pp. 1–4). New York: Wiley.

Broadbent, G., Bunt, R., & Llorens, T. (Eds.). (1980). *Meaning and behavior in the built environment.* New York: Wiley.

Broadbent, G., Bunt, R., & Jencks, C. (Eds.). (1980). *Signs, symbols and architecture.* New York: Wiley.

Castex, J., Depaule, J. C., & Panerai, P. (1978). Essai sur les structures syntaxiques de l'espace architectural. *Notes méthodologiques en architecture et en urbanisme* (Sémiotique de l'espace), *7*, 101–155.

Castex, J., & Panerai, P. (1974). Structures de l'espace architectural. *Notes méthodologiques en architecture et en urbanisme* (Sémiotique de l'espace), *3/4*, 39–63.

Castex, J., & Panerai, P. (1979). Structures de l'espace architectural. In *Sémiotique de l'espace* (pp. 61–93). Editions Denoël/Gonthier.

Choay, F. (1970). Urbanism and semiology. In C. Jencks & G. Baird (Eds.), *Meaning in architecture* (pp. 26–37). London: Barrie & Rockliff.

Choay, F. (1970–1971). Remarques à propos de sémiologie urbaine. *L'architecture d'aujourd'hui* (La ville), *153*, 9–10.

Choay, F. (1973). Figures d'un discours méconnu. *Critique* (L'urbain et l'architecture), *311*, 293–317.

Duncan, J. S. (1985). The house as symbol of social structure: Notes on the language of objects among collectivistic groups. In I. Altman & C. M. Werner (Eds.), *Home environments* (pp. 133–151). New York: Plenum.

Eco, U. (1968). *La struttura assente.* Milano: Bompiani.

Eco, U. (1972). A componential analysis of the architectural sign/column. *Semiotica, 5*(2), 97–117.

Eco, V. (1980). A componential analysis of the architectural sign /column/. In G. Broadbent, R. Bunt, & C. Jencks (Eds.), *Signs, symbols, and architecture* (pp. 213–232). New York: Wiley.

Fauque, R. (1973). Pour une nouvelle approach sémiologique de la ville. *Espace et sociétés, 9*, 15–27.

Fauque, R. (1979). Le discours de la ville. In S. Chatman, U. Eco, & J. M. Klinkenberg (Eds.), *A semiotic landscape* (pp. 918–923). Den Hague: Mouton.

Franke, J. (1969). Stadtbild: Zum Erleben der Wohnumgebung. *Stadtbauwelt, 24*, 292–295.

Fusco, R. de (with M. L. Scalvini) (1960). Significanti e significanti nella rotonda palladiana. *Op. cit. Selezione della critica d'arte conteni poranea, 16*, 5–23.

Fusco, R. de (1967). *Architettura come mass-medium.* Bari: Dedalo.

Gamberini, I. (1953). *Per una analisi degli elementi dell'architettura.* Firenze: Editrice Universitaria.

Gamberini, I. (1959). *Gli elementi dell'architettura come 'parole' del linguaggio architettonico.* Firenze: Coppini.

Gamberini, I. (1961). *Analisi degli elementi costitutivi dell'architettura.* Firenze: Coppini.

Garroni, E. (1964). *La crisi semantica della arti.* Roma: Officina.

Garroni, E. (1972). *Progetto di semiotica.* Bari: Laterza.

Gibson, J. J. (1979). *The ecological approach to visual perception.* Boston: Houghton Mifflin.

Gibson, J. J. (1982). Notes on affordances. In E. Reed & R. Jones (Eds.), *Reasons for realism: Selected essays of James J. Gibson* (pp. 401–418). Hillsdale, NJ: Erlbaum.

Gottdiener, M. (1986). Urban culture. In T. A. Sebeok (Ed.), *Encyclopedic dictionary of semiotics* (Vol. 2, pp. 1141–1145). Amsterdam: Mouton de Gruyter.

Greimas, A. (1974). Pour une sémiotique topologique. *Notes méthodologiques en architecture et en urbanisme* (Sémiotique de l'espace). *3/4*, 1–21.

Greimas, A. (1979). Pour une sémiotique topologique. In *Sémiotique de l'espace* (pp. 11–43). Paris: Editions Denoel/Gonthier.

Environmental Meaning 265

Groat, L. (1982). Meaning in post-modern architecture: An examination using the multiple sorting tool. *Journal of Environmental Psychology, 2,* 3–22.
Groat, L., & Canter, D. (1979). Does post-modernism communicate? *Progressive Architecture, 12,* 84–87.
Haeckel, E. (1866) *Generelle Morphologie der Organismen,* Vol. 2: *Allgemeine Entrocklungsgeschichte der Organismen.* Berlin: Reimer.
Hall, E. T. (1966). *The hidden dimension.* New York: Doubleday.
Hammad, M. (1979). Sémiotique de l'espace et sémiotique de l'architecture. In S. Chatman, U. Eco, & J.-M. Klinkenberg (Eds.), *A semiotic landscape* (pp. 925–929). Paris: Mouton.
Harrison, D., & Howard, W. A. (1972). The role of meaning in the urban image. *Environment and Behavior, 4.*
Hershberger, R. G. (1972). Toward a set of semantic scales to measure the meaning of architectural environments. In W. J. Mitchell (Ed.), *Environmental design: Research and practice, Vol. 3* (pp. 6-4-1–6-4-10). Stroudsburg, PA: Dowden, Hutchinson & Ross.
Hillier, B., & Leaman, A. (1973). The man–environment paradigm and its paradoxes. *Architectural Design, 8,* 507–511.
Hillier, B., & Leaman, A. (1975). The architecture of architecture. In D. Hawkes (Ed.), *Models and systems in architecture and building* (pp. 5–23). London: Construction Press.
Hillier, B., Leaman, A., Stansall, P., & Bedford, M. (1976). Space syntax. *Environment and Planning B, 3,* 147–185.
Hjelmslev, L. (1968–1971). *Prolégomènes à une théorie du langage.* Paris: Les Editions de Minuit.
Honikman, B. (1973). Personal construct theory and environmental evaluation. In G. Broadbent, R. Bunt, & T. Llorens (Eds.), *Meaning and behavior in the built environment* (pp. 79–91). New York: Wiley.
Hunter, A. (1987). The symbolic ecology of suburbia. In I. Altman & A. Wandersman (Eds.), *Neighborhood and community environments* (pp. 191–221). New York: Plenum.
Jakobson, R. (1960). Closing statement: Linguistics and poetics. In T. A. Sebeok (Ed.), *Style in language* (pp. 350–377). Cambridge, MA: MIT Press.
Jencks, C. (Ed.). (1969). *Meaning in architecture.* London: Barrie & Rockliff.
Jencks, C. (1980). The architectural sign. In G. Broadbent, R. Bunt, & C. Jencks (Eds.), *Signs, symbols, and architecture* (pp. 71–118). New York: Wiley.
Kasmar, J. (1970). The development of a usable lexicon of environmental descriptors. *Environment and Behavior, 2,* 153–169.
Kelly, G. A. (1955). *The psychology of personal constructs.* New York: Norton.
Koenig, G. K. (1964). *Analisi del linguaggio architettonico.* Firenze: Liberia Editrice Fiorentina.
Koenig, G. K. (1970). *Architettura e communicazione.* Preceduta da elementi di analisi del linguaggio architettonico. Firenze: Liberia Editrice Fiorentina.
Koffka, K. (1935). *Principles of gestalt psychology.* New York: Harcourt Brace.
Krampen, M. (1971). Das Messen von Bedeutung in Architektur, Stadtplanung und Design. Teil 1: Das Polaritätsprofil als Meßinstrument. *Werk, 1,* 57–60.
Krampen, M. (1974). A possible analogy between (psycho-) linguistic and architectural measurement—the type-token ration (TTR). In D. Canter & T. Lee (Eds.), *Psychology and the built environment* (pp. 87–95). London: Architectural Press.
Krampen, M. (1979a). *Meaning in the urban environment.* London: Pion.
Krampen, M. (1979b). Survey on current work in semiology of architecture. In S. Chatman, U. Eco, & J.-M. Klinkenberg (Eds.), *A semiotic landscape* (pp. 169–194). The Hague: Mouton.
Krampen, M. (1980). The correlation of "objective" facade measurements with subjective

facade ratings. In G. Broadbent, R. Bunt, & T. Llovens (Eds), *Meaning and behavior in the built environment*. (pp. 61–78). New York: Wiley.

Krampen, M. (1989). Semiotics in architecture and industrial product design. *Design Issues, 5*(2), 124–140.

Lagopoulos, A.-P. (1978). Analyse sémiotique de l'agglomération européenne pré-capitaliste. *Semiotica, 23*(1/2), 99–164

Lagopoulos, A.-P. (1986). Settlement space. In T. A. Sebeok (Ed.), *Encyclopedic dictionary of semiotics* (Vol. 2, pp. 924–936). Amsterdam: Mouton de Gruyter.

Lawrence, R. J. (1985). A more humane history of homes: Research method and application. In I. Altman & C. M. Werner (Eds.), *Home environments* (pp. 113–132). New York: Plenum.

Ledrut, R. (1970). L'image de la ville. *Espaces et sociétés, 1*, 93–106.

Ledrut, R. (1973). Parole et silence de la ville. *Espaces et sociétés, 9*, 3–14.

Lewin, K. (1926). Untersuhungen zur Handlungs—und Affekt—Psychologie, I, II. *Psychologische Forschung, 7*, 294–385.

Lingoes, J. C. (1973). *The Guttman-Lingoes nonmetric program series*. Ann Arbor, MI: Mathesis Press.

Lynch, K. (1960). *The image of the city*. Cambridge, MA: MIT Press.

Miller, G. A. (1967). Psycholinguistic approaches to the study of communication. In D. L. Arm (Ed.), *Journeys in science* (pp. 22–73). Albuquerque: University of New Mexico Press.

Miller, G. (1969). A psychological method to investigate verbal meaning. *Journal of Mathematical Psychology, 6*, 169–191.

Moore, G. T. (1979). Knowing about environmental knowing. *Environment and Behavior, 11*, 33–70.

Moore, G. T., & Golledge, R. G. (Eds.). (1976). *Environmental knowing: Theory, research, and methods*. New York: Van Nostrand Reinhold.

Morris, C. W. (1964). *Signification and significance*. Cambridge, MA: MIT Press.

Mukarovsky, J. (1978). On the problem of function in architecture. In J. Burbank & P. Steiner (Eds.), *Structure, sign, and function: Selected essays by Jan Mukarovsky* (pp. 236–250). New Haven: Yale University Press.

Nöth, W. (1985). (Ed.). Architektur. In *Handbuch der Semiotik* (pp. 400–408). Stuttgart: J. B. Metzlersche Verlagbuchhandlung.

Osgood, C. E., Suci, G. J., & Tannenbaum, P. H. (1957). *The measurement of meaning*. Urbana: University of Illinois Press.

Ostrowetsky, S., & Bordreuil, S. (1979). Sociologie et sémiotique. In S. Chapman, U. Eco, & J.-M. Klinkenberg (Eds.), *A semiotic landscape* (pp. 956–959). Den Hague: Mouton.

Piaget, J., & Tuhelder, B. (1948). *La représentation de l'espace chez l'enfant*. Paris: Presses universitaire de France. (English translation by F. J. Langdon & J. L. Lunzre: *The child's conception of space*. London: Routledge & Kegan Paul, 1956.)

Pinxten, R. (1974). Emicism and how to avoid a paradox. *Communication and Cognition, 7*(3/4), 315–333.

Pinxten, R. (1976). Epistemic universals: A contribution to cognitive anthropology. In R. Pinxten (Ed.), *Universalism and relativism in language and thought* (pp. 117–175). Den Hague: Mouton.

Pinxten, R. (1977). Descriptive semantics and cognitive anthropology: In search for a new model. *Communication and Cognition, 10*(3/4), 89–106.

Posner, R. (1986). Zur Systematik der beschréibung verbaler und nonverbaler kommunikation. In, H. G. Bosshardt (Ed.), *Perspektiven auf sprache* (pp. 267–313). Walter de Gruyter.

Preziosi, D. (1979a). *Architecture, language, and meaning*. Den Hague: Mouton.
Preziosi, D. (1979b). *The semiotics of the built environment: An introduction to architectonic analysis*. Bloomington: Indiana University Press.
Preziosi, D. (1983a). The network of architectonic signs. In T. Borbé (Ed.), *Semiotics unfolding* (Vol. 3, pp. 1343–1349). Den Hague: Mouton.
Preziosi, D. (1983b). Minoan architectural design. Berlin: Mouton.
Preziosi, D. (1986). Architecture. In T. A. Sebeok (Ed.), *Encyclopedic dictionary of semiotics* (Vol. 1, pp. 44–50). Amsterdam: Mouton de Gruyter.
Proshansky, H. M., Ittelson, W. H., & Rivlin, L. G. (1970). *Environmental psychology: Man and his physical setting*. New York: Holt, Rinehart & Winston.
Rapoport, A. (1982). *The meaning of the built environment: A nonverbal communication approach*. Beverly Hills, CA: Sage.
Renier, A. (1979). Nature et lecture de l'espace architectural. In *Sémiotique de l'espace* (pp. 44–59). Paris: Editions Denoël/Gonthier.
Rossi-Landi, F. (1968). *Il linguaggio come lavoro e come mercato*. Milano: Bompiani.
Rossi-Landi, (1972). Omologia della riproduzione sociale. *Ideologia, 16/17*, 43–103.
Rossi-Landi, F. (1975). *Linguistics and economics*. Den Hague: Mouton.
Ruesch, J. & Kees, W. (1970). Function and meaning in the physical environment. In H. M. Proshansky, W. H. Ittelson, & L. G. Rivlin (Eds.), *Environmental psychology: Man and his physical setting* (pp. 141–153). New York: Holt, Rinehart & Winston.
Sanchez-Robles, C. (1980). The social conceptualization of home. In G. Broadbent, R. Bunt, & T. Llorens (Eds.), *Meaning and behavior in the built environment* (pp. 113–133). New York: Wiley.
Scalvini, M. L. (1968). Per una teoria dell'architettura. *Op. Cit.: Selezione della critica d'arte contemporanea, 13*, 30–44.
Scalvini, M. L. (1975). *L'architettura come semiotica connotative*. Milano: Bompiani.
Scalvini, M. L. (1979). A semiotic approach to architectural criticism. In S. Chatman, U. Eco, & J.-M. Klinkenberg (Eds.), *A semiotic landscape* (pp. 965–969). New York: Mouton.
Scalvini, M. L. (1980). Structural linguistics versus the semiotics of literature: Alternative models for architectural criticism. In G. Broadbent, R. Bunt, & C. Jencks (Eds.), *Signs, symbols, and architecture* (pp. 411–420). New York: Wiley.
Sebeok, T. A. (Ed.). (1986). *Encyclopedic dictionary of semiotics* (Vols. 1–3). Den Hague: Mouton.
Sommer, R. (1965). The significance of space. *Journal of the American Institute of Architecture*, May, 63–65.
Stringer, P. (1980). The meaning of alternative future environments for individuals. In G. Broadbent, R. Bunt, & T. Llorens (Eds.), *Meaning and behavior in the built environment* (pp. 93–111). New York: Wiley.
Stokols, D., & Altman, I. (Eds.). (1987). *Handbook of environmental psychology*. New York: Wiley.
Uexküll J. von (Ed.). (1913). Tierwelt oder Tierseele. In *Bausteine zu einer biologischen Weltanschauung* (pp. 77–100). München: Bruckmann.
Walther, E. (1974). *Allgemeine Zeichenlehre. Einfuhrung in die Grundlagen der Semiotik*. Stuttgart: Deutsch Verlagsanstalt.
Wapner, S., Kaplan, B., & Cohen, S. B. (1980). An organismic-developmental perspective for understanding transactions of men and environments. In G. Broadbent, R. Bunt, & T. Lloren (Eds.), *Meaning and behavior in the built environment* (pp. 223–255). New York: Wiley.
Warren, W. J. (1984). Perceiving affordances: Visual guidance of stair climbing. *Journal of Experimental Psychology: Human Perception and Performance, 10*, 683–703.

Wohlwill, J. F. (1976). Environmental aesthetics: The environment as a source of affect. In I. Altman & J. F. Wohlwill (Eds.), *Human behavior and environment* (Vol. 1, pp. 37–86). New York: Plenum.

Wohlwill, J. F. (1977). Environmental psychology: An overview. In B. B. Wolman (Ed.), *International encyclopedia of psychiatry, psychology, psychoanalysis & neurology* (Vol. 4, pp. 338–341). New York: Aesculapius.

Wolman, B. B. (Ed.). (1977). *International encyclopedia of psychiatry, psychology, psychoanalysis & neurology* (Vol. 4). New York: Aesculapius.

Zube, E. H. (1976). Perception of landscape and land use. In I. Altman & J. F. Wohlwill (Eds.), *Human behavior and environment* (Vol. 1, pp. 87–121). New York: Plenum.

V

ADVANCES IN RESEARCH
UTILIZATION

8

Participatory and Action Research Methods

BEN WISNER, DAVID STEA, and SONIA KRUKS

ACTION AND PARTICIPATION

ACTION RESEARCH: ORIGINS AND PRELIMINARY DEFINITION

The term "action research" is associated in the USA with Kurt Lewin, who published extensively on action research from the 1940s onward and gave these methods their conceptual form. To Lewin (1946), action research was

> comparative research on the conditions and effects of various forms of social action and research leading to social action . . . . [it is] a big spiral of steps, each of which is composed of a circle of planning, action, and fact-finding about the result of the action. [Action, research, and training form] a triangle that should be kept together for the sake of any of the corners. (pp. 202–211)

In Lewin's definition and description, it is clear that an early objective of action research was the *solution of social problems* rather than the mere compilation of scientific data. Second, action research was con-

Ben Wisner • School of Social Science, Hampshire College, Amherst, Massachusetts 01002. **David Stea** • International Center for Built Environment, Santa Fe, New Mexico 87131. Present address of D.S.: Universidad National Autonoma de Mexico, Mexico D.F., Mexico. **Sonia Kruks** • Department of Government, Oberlin College, Oberlin, Ohio 44074.

ceived as an *iterative* process. Unlike much social science research, it might have to be repeated again and again to be efficacious. Third, emphasis was placed upon *training* social scientists to handle not just the usual forms of scientific research but also to translate such research into practice in the field. The simultaneity of research and praxis is therefore an essential element of this approach. Fourth, *communication*—especially intercultural communication—between planners and clients was an important element in such field training/research/praxis.

One is tempted to compare action research as practiced (perhaps too rarely) in psychology and sociology with "participant observation" in anthropology. The critical difference is that, unlike the action researcher, the participant observer is supposed *not* to be a change agent and is expected to influence the ambient situation as little as possible: no praxis is involved. The participant observer is thus involved and detached at the same time—no small achievement. The element of praxis is distinctly absent from orthodox concepts of participant observation.

Today it is recognized by action researchers that in the give-and-take of praxis, the lay person must be as much of a participant as the researcher (Gardner, 1974; Weisman, 1983). However, there seems to be an almost indefinite number of ways in which citizens can be approached as participants and almost as many ways of conceptualizing this relationship. Much of what we review in this chapter centers on just what participation is, can be, or should be.

Examples of participatory action research include the world-renowned work of Turner in the field of self-built housing (1984) as well as other efforts in housing and neighborhood design (Churchman, 1987; Francis, 1979; Nicholson, 1974; Pyatok & Weber, 1978). Examples in resource management have included a community-based natural resource inventory (Field, 1986), evaluation of rice varieties (Richards, 1985), and community design of water catchment systems (Guggenheim & Fanale, 1975). Women's participation has been used in designing fuel-efficient cooking stoves (Aprovecho, 1980). Campbell (1987) employed participatory methods in the evaluation of government hoof-and-mouth disease control programs. Other applications to evaluation include an attempt by Oxfam-America to evaluate its grass-roots projects in southern Africa (Kalyalya, Mhlanga, Semboja, & Seidman, 1988) and postoccupancy evaluation studies (Kantrowitz & Nordhaus, 1980). There have been many other applications in the fields of public health, social welfare, and farming systems research.

There are close resemblances between Lewin's view of action research and intellectual movements in the 1920s, 1930s, and 1940s in Asia and Latin America. This includes the "methods of work" advocated by

Mao Tse-tung during the Chinese revolution, when cadres were supposed to study the concrete material problems of daily life together with peasants in the liberated zones (Mao Tse-tung, 1967), and the methods employed by Gandhian intellectuals in the 1920s and 1930s, who fanned out to many flash points of nationalist agitation to document with the people economic and social grievances (Gandhi, 1952; cf. Bhaduri & Rahman, 1982; Manandhar, 1986; Rahman, 1984). In Latin America, the roots of action research lie in the more general dialogical process of "conscientization" (Freire, 1973). In this work there is an emphasis on the study by people of their concrete living situation (*estudo do meio*). It was a very small step, then, to the common use of *investigación-acción* (action research) by grass-roots rural groups in many parts of Latin America to articulate more clearly to themselves their economic, social, and political options in the face of specific material expressions of poverty and oppression (Brandao, 1984; Vio Grossi, Gianotten, & de Wit, 1981). Somewhat later, in Africa, the decolonialization of the formal school system led to similar ideas of "relevant" research by the school-centered rural community and its own material problems, best expressed in former Tanzanian president Nyerere's well-known essay "Education for Self-Reliance" (Nyerere, 1968).

Internationally the approach of "research for the people/research by the people" has been taken up by many kinds of professionals linked in various direct and indirect ways to the academic fields of adult, informal, and continuing education (Dubell, Erasmie, & De Vries, 1981) and supported by international organizations such as the International Council for Adult Education (Hall, Gillette, & Tandon, 1982). While adult education institutions have provided an early and continuous "home port" for action research, the evolution of participatory methods in planning cannot be divorced from parallel developments in public health (Morley, Rohde, & Williams, 1983; Newell, 1975), farm systems research or "rapid rural appraisal" (Blaut, 1967; Longhurst, 1981; Maxwell, 1984; Richards, 1985), and applied ethnolinguistics and anthropology (Brokensha, Warren, & Werner, 1980).

Work along participatory lines emerging from within the Third World has often been bound up with cultural nationalism, which analyzed the relations of dominance and dependency that had been created by the transfer of technology from colonial centers of science and industry (Mbilinyi, Vuerla, Kassam, & Masisi, 1982; UNESCO, 1981). Elsewhere in the Third World efforts such as that organized by the United Nations University have focused on increasing the direct dialogue between rural people and university-based researchers in, for instance, Ethiopia, the Philippines, and Mexico (Herrera, 1984).

Participation and Participatory Research

The term action research implies an *actor*, the researcher. In participation, emphasis is placed upon the *participants* as the primary actors: thus, participation research may not be the same as action research in the sense that the researcher may not be the primary investigator or, as in the case of "advocacy planning," the advocate may play only a catalyzing/synthesizing role (Davidoff, 1965; Peattie, 1968; Wisner, 1970). It is thus possible to have action research that is not participatory and participation research that is nonaction. In many cases, however, the overlap is considerable: most participation efforts have an action component, and much action research is participatory. Thus, we can consider at least three aspects of participation: (1) as action/praxis, (2) as communication, and (3) as research.

Participation as *action/praxis* stresses the process of participation itself (Freire, 1973; Swantz, 1975), and its effects upon both the participants and the outcome of the process. In some cases, especially in the Two-Thirds World, this emphasis on the effect on people and their consciousness means that the established mechanisms of "citizen participation" and official decision making may be called into question where certain groups (e.g., women, slum dwellers, marginal farmers) are excluded (Vio Grossi, 1981). Such a participation process does not necessarily harmonize divergent social and economic interests, but can be part of a conflictual process of negotiating such interests (Huizer, 1984).

Participation as *communication* emphasizes the interaction between professionals and the citizens whom they serve and focuses more narrowly on the design product resulting from such communication (or lack of it). For instance, environmental designers and physical planners are concerned with translating verbal input into graphic product, and with the translation of the resulting feedback. Often quite simplistic and mechanistic models of "communication" are accepted uncritically by such design workers, although their "products" may be substantially "better" (in various ways) than those produced without participation.

Participation as *research* emphasizes the systematic documentation of the process and product, with the eventual end of formulating frameworks, models, and theories of participation, as well as evaluating its efficacy. There is some debate, however, over what most needs to be documented: actions of the participants, the process, or the outcome. In environmental design and planning, for example, there is additional concern about the value of participation to the products of design. The question is asked whether participation actually "works" as a technique contributing to the design of enhanced environments (Wandersman, 1987).

By contrast, in the areas of public health, agriculture, and resource management, there are fewer doubts expressed about the design outcomes of participatory work. It is becoming recognized in these fields that often this is the only way to achieve sustainable solutions (Chambers, 1983). This may have to do with the possibly more complex nature of epidemiological and agro-ecological system when compared to design of architectural systems. The contrasting response of "the professionals" to participatory methods could also mirror a difference in professional training and power structure (Chambers, 1985) between physical and social planners.

Quite a number of studies have examined specific techniques of participation. These include surveys (e.g., Becker, 1977); discussion groups, sociodrama, and traditional forms of storytelling (Bhasin, 1978; Hilton, 1983; Mwansa & Kidd, 1982; Peattie, 1970); games and related simulations (Dandekar & Feldt, 1984; McKenchnie, 1977; Sanoff, 1979); and the use of environmental models (Bentz, 1988; Jacome, 1988; Lawrence, 1983, 1988; Sanchez, Cronick & Wiesenfeld, 1988; Simon, 1988; Stea, 1980, 1981a, 1981b, 1982, 1984, 1988). R. Kaplan (1987) has compared "high-articulation" (finished architectural) models and "low-articulation" (simplified, schematic) models in participation, concluding:

> lack of design training does not interfere with citizens' ability to make meaningful judgments of alternative site arrangements early in the design process. Expensive models are not needed to engage them in the process and to elicit feedback from them to incorporate in final decisions. (p. 101)

The "new technologies" have had their impact, as well, and the development of interactive computer programs has enabled the use of computers in participatory architectural design (e.g., Bollinger, 1986; Brown & Novitski, 1986; Negroponte, 1970, 1975). Writers on this subject have pointed out the advantages and disadvantages of computer-aided participation. It is possible for the user to obtain instantaneous feedback on the consequences of a decision, for example, but participation of this sort is at present individual rather than social, and also, quite obviously, restricted to situations where computers and appropriate programs are available (which is not the case in many developing countries).

FRAMEWORKS FOR PARTICIPATION

A Social Framework

Early efforts to document participation were primarily compilations of case studies. More recent work has begun to examine the problem of systematizing participation research, resulting in the generation of models or frameworks for environmental design rather than true theories (e.g. Cashdan, Fahle, Schwartz, & Stein, 1979; S. Kaplan, 1977; Palmer, 1985; Patricios, 1979; Rand, Heath, & Wilde, 1977; Smith, 1978; Sommer, 1977). Wandersman (1987) has provided a list of significant questions addressed by his colleagues and himself in their research, some of which are:

1. Who participates, who does not participate, and why?
2. How does the interaction of the person and the situation influence participation?
3. What is a sense of community and what are its consequences?
4. What are the characteristics of organizations that are active and successful versus those that become inactive?
5. What cross-culture comparisons are appropriate to participation in community development?
6. With regard to knowledge dissemination and knowledge utilization, what is the relationship between scientists and citizens?

Theorists in political science and sociology have provided ways of putting such specific questions into a broader context. Thus, "participation" has been situated in democratic theory (Barber, 1984; Mansbridge, 1980; Pateman, 1964) and within the theory of organizations and bureaucracies (Abrahamsson, 1977; Himmelstrand, 1981). The question of power and control of one social group over another is often judged to be central to the understanding of participation and obstacles to it (Gran, 1983).

These more philosophical reflections on participation often draw a critical distinction between participation initiated or utilized by external power holders to achieve their own ends through mobilizing local community labor or other material support and participation initiated locally as part of an ongoing struggle by people to control their own lives. The contrast is between *instrumental* and *transformative* participation (Kruks, 1983). Advocates of "instrumental" participation see it as "a Good Thing primarily because it is effective or efficient"; whereas the "transformative" variety appears to its advocates "a Good Thing because it is seen to facilitate changing social consciousness and, ultimately, to facilitate changing the present unjust distributions of wealth and power in many societies" (Kruks, 1983, p. 3).

Many authors emphasize that participation is especially difficult in situations where a particular group has been systematically excluded from power for long periods of time. Blaikie and Brookfield (1987) note that there is often an overlap and mutual reinforcement of political,

TABLE 1. Some Differences between Community and Traditional
Design Practices

Community design[a]	Traditional design[b]
Small scale	Large scale
Local	National/international
Appropriate technology	High technology
Human oriented	Corporate or institutionally oriented
Client redefined to include users	Single-client oriented
Process and action oriented	Building and project oriented
Concerned with meaning and ornament	Concerned with style and context
Low cost	High cost
Bottom-up design approach	Top-down design approach
Inclusive	Exclusive
Democratic	Authoritarian

[a]As practiced by community landscape designers.
[b]As practiced by larger architectural planning firms.

economic, and ecological marginality of such groups (e.g., nomads in
Africa, "tribals" in India, etc.). Poor women, as a group, experience
many obstacles to participation (Lamming, 1980). In rural Africa, for
instance, even where institutional efforts have been made to mobilize
women, both young women and widows as well as divorcees may still
be excluded (Kruks & Wisner, 1984; Feldman, 1984). Indeed, there is
even a great deal of confusion about how to *conceptualize* women's exclusion from economic and political participation, let alone what to do
about it (Kruks, Rapp, & Young, 1989; Rogers, 1980).

The frameworks discussed so far address few of these more troub-
ling problems, but they do offer a way of understanding how attempts at
participation—albeit within such serious limitations—differ from conventional planning methods. For instance, Francis (1979) has tabulated
the contrasts between participatory community design and the traditional design practice, process, and products of large architecture, landscape architecture, and urban design firms, providing a framework for
comparison and evaluation (see Table 1). With modifications, this framework should be generally applicable to other planning realms as well.

Wandersman (1987) has provided a different kind of framework
using a flow diagram to interrelate the questions he put forth, listed
earlier (see Figure 1). The framework first asks "who participates?" then
goes on to consider five parameters of participation which, through
"mediators," produce effects. The cycle is completed through a feedback

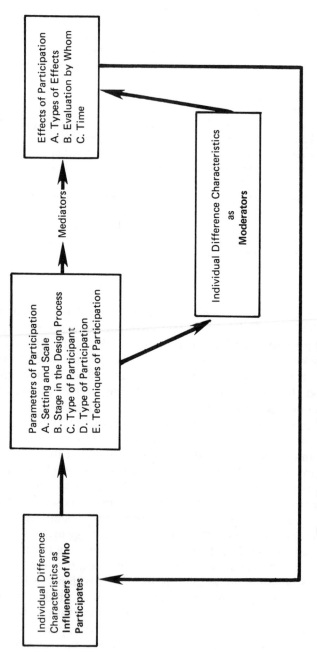

FIGURE 1. A conceptual framework of user participation in planning environments.

loop linking effects to the "who participates" question, once again. Wandersman is also concerned with individual differences as moderators, and this plays a role, as well, in his framework.

A COMMUNICATION FRAMEWORK

One central aspect of participation is communication: participants in a design/planning process, for example, must communicate with design/planning professionals, and vice-versa. It is almost tautological to state that without communication, there is no participation. Hence, a significant question concerning the participation process is: what facilitates communication and what hampers it?

Communication theory was formulated nearly 40 years ago (Shannon & Weaver, 1949) to describe characteristics of electronic communications systems. Although the underlying model is mechanistic—involving relations among "transmitters," "signals," and "receivers"—this theory, in one modification or another, has since underpinned virtually all applied work on communication among planners and "extension" personnel. For instance, Palmer (1985) asserted that relatively slight modifications are needed to enable the model to describe face-to-face interactions among humans. Then, the source becomes the human brain; the transmitter is the vocal apparatus (for verbal material) or the human hand (for graphic information); the signals are sound waves (verbal) or light waves (graphic) reaching the receiver's ear or eye and being transmitted to the destination (the receiver's brain). In a typical design application, Stea (1984) proposed further simplification to represent the typical participation sessions.

In the diagram in Figure 2, citizens convey information verbally, while the professionals process this verbal information into graphic form, to which the participants are expected to respond verbally. Recalling the Shannon–Weaver model, we note the presence of a "noise source" that imposes "noise" on the "signal." The noise in this case occurs at several stages, involving the translation of visual images in the head of a citizen into a verbal message, and the reception of that verbal message by a professional who must in turn translate the verbal material into a mental image. This mental image is then further translated into a graphic output that is supposed to be understood by the participating citizen, who is then expected to provide verbal feedback. If noise is taken here to represent opportunities for distortion of messages, then there are at least four such opportunities, indicated by numbers on the diagram in Figure 2.

A diagram such as this stresses that one goal of participation should

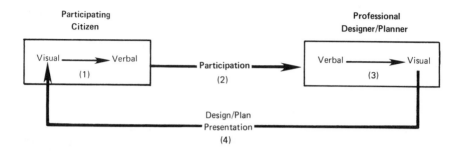

FIGURE 2. Simplified flow of transduced information/conventional participation process.

be a reduction in the number of opportunities for distortions of information and consequent misunderstandings. This is especially true when participants and professionals are from different socioeconomic classes or where linguistic, cognitive, and other cultural differences exist. Sources of noise, then, include (1) the participation environment (community meetings, face-to-face interviews, etc., and the places in which these occur), (2) the medium of communication (verbal, graphic, games/simulations, etc.), (3) professional subcultures and their jargon, (4) socioeconomic class differences, (5) cultural differences, and (6) language barriers.

Communication has, of course, been addressed from strikingly different intellectual traditions, and many planners are currently becoming aware of the limits of such models derived from electrical engineering and cybernetics. While the models just discussed presuppose relatively harmonious or consensual social orders, not all students of communication share those assumptions. For example, radical and neo-Marxist perspectives on "ideology" suggest that what the above models conceptualize as noise in communication could actually be intended by dominant classes in order to confuse, mystify, and demobilize the oppressed (Habermas, 1975; Marx & Engels, 1970; Thompson, 1984). In development studies the notion of "voice" has arisen to express the possibility that power in society can render certain groups "silent" (Freire, 1973; Schaffer & Lamb, 1981), while feminists have pointed out that in many cases it is women who are rendered "invisible" and made silent. Even more radical are contemporary deconstructionist views of language as itself a vehicle of oppression and communication as never "transparent" (Derrida, 1978), while for Foucault (1980) the inseparability of "power/knowledge" precludes all possibility of "neutral" or nonpower laden relations between researcher/designers and client/ participants.

CASE STUDIES

While earlier action research focused upon the participation of the researcher in the ongoing activities of a subject population, the current emphasis, mirrored in the following five cases, is upon the attempt at *coequal* participation of citizen and researcher alike. These case studies illustrate participation as praxis, communication, and research. All of them are oriented toward community-centered planning/action as opposed to previously accepted top-down approaches, juxtaposed in Table 1 above. In various ways, each case explores the dynamics of the process of participation itself, set out schematically in Figure 1 above, although Case I was far more self-conscious on this point. Similarly, all cases present various solutions to the problem presented by communication across the gaps created by class, ethnicity, age, and gender. Here again, there are differing degrees of self-consciousness. In Case II, for instance, the absence of a women's participation was seen as a grave but nearly unavoidable limitation; while in Case III the more or less uncritical acceptance of a modified Shannon–Weaver model (Figure 2 above) meant that communication was considered unproblematic.

Finally, in all five studies the distinction made earlier between instrumental and transformative participation is revealed. Both "design product" and "process" feature as goals of participation, but the balance between product and process varies from case to case. For instance, in cases I and II, development was seen as not always harmonizing diverse social goals, but in some circumstances entailing arbitration, negotiation, and mediation of conflict. Thus, the full potential of participatory approaches is revealed only when one views the process through both the lens of psychology/communications models and social theories involving contradictions.

Case I: Casalta II—Reconstruction following a Landslide

Several years ago, the inhabitants of an irregular settlement called *Nazareno*, in Caracas, Venezuela, were rendered homeless by a landslide. Refusing to move into barracks intended for their temporary shelter, they broke into a nearby school. Later, community members requested the donation of land on which to build a new and much more secure settlement, and asked for the collaboration of members of the teaching staff and students at the School of Architecture of the Central University of Venezuela. The University responded by working with community members to form a joint organization called *Escuela Popular de Arquitectura* (People's School of Architecture). Because problems of intercultural communication were seen as affecting the interaction between two widely divergent socioeconomic groupings, the involvement of social psychologists was requested by the newly formed *Escuela Popular de Arquitectura*.

The social psychologists uncovered, first, the concern of community leaders with the passivity of many community members, whose initial concepts of what constitutes participation were very diverse. A need was thus recognized for a variety of criteria concerning what participation really is. Second, there appeared to be a strong desire to organize motivation systems and mechanisms of participation for people with limited time and resources. The *Escuela Popular de Arquitectura* team, augmented with the social psychology group, worked together very closely, eventually incorporating an outside expert on *autoconstrucción* (self-help construction).

Among the conclusions drawn by the social psychologists on the team concerning the nature of participation itself, the following are worthy of special note (Sanchez *et al.*, 1988):

1. Participation is not a unitary construct, but all reasonable definitions involve the distribution of *power*. In Casalta II, people were not offered power by government officials; rather, and this is one possible reason for the success of the project, they *took power* for themselves.

2. Casalta II was characterized by *active* participation: the community was clearly dependent upon the government for construction materials, but its people demanded that government officials help them, as citizens in need. When they encountered resistance, they used newspapers and radio to make public their complaints concerning the treatment they received.

The initiative had come from within the community and its relation to traditional holders of power was changed in the process. Thus, participation in this case must be called transformative and not merely instrumental.

The social psychologists also noted three determinants of such active (or what we would term "transformative") participation. These appeared to the researchers like a three-legged stool, resting upon community *organization, solidarity* within the community, and the selection of *appropriate forms* or means of participation (the question, yet again, of technique).

CASE II: LOCALLY IMPROVED GRAIN STORAGE IN A TANZANIAN VILLAGE

In 1976 the Republic of Tanzania and the Dutch Overseas Development Ministry took an innovative approach to the design of grain stor-

FIGURE 3. Final summary problem code, "Enemies of the Farmer—*Adui za Wakulima*."

age improvements. An interdisciplinary team was created. Rather than apply state-of-the-art designs for low-cost grain storage, the team was charged to develop with villagers an appropriate range of improved designs (Mduma, 1982; Wisner *et al.*, 1979).

The team worked for three months in a village in Morogoro District at the intersection of the Uluguru Mountains and the plains that drain into the Rufiji river. Village-level discussions conducted during the previous year had identified storage as a priority in Morogoro District generally, and the village of Bwakira Chini had, in particular, already lodged a request with the government for assistance with storage.

Work in this village proceeded in three stages. First, a counterpart team of lay experts from the village were elected by the village council. They and the outside team then set about making an inventory of all existing storage systems in the village. Everyone was surprised by the diversity of designs already existing in a village of 270 households. During this first stage, discussions were held in four sites in the village that elicited from people their problems and experience with storage. A Tanzanian artist in the outside team rendered many of the things said in these sessions as large water color posters that were presented to the group the next day (see Figure 3). Such "codes" deepened discussion and allowed the outside team to determine whether they had "gotten the point."

Stage two was that of design. Again the artist presented numerous water color versions of alternative designs. These were criticized. Cost, exposure to specific weather conditions, vulnerability to theft, strength of locally available building materials were issues raised by villagers. Stage one had produced two major items of village consensus. An existing outside grain crib on stilts (called a *dungu*) could be protected from rats and would be better ventilated. A common indoors system of storage in the roof over the cooking fire (called a *dari*) could not be protected from rats but was safe from theft and envy and took advantage of insecticidal properties of smoke from the cooking fire and its drying effects. A major conceptual breakthrough for outsiders and villagers alike was the realization that a modified Nigerian prototype, the best of both *dungu* and *dari*, could be built at low cost.

During stage three, several actual prototypes were built. The physical properties of building materials were tested. The viability of innovations such as recycling old kerosene tins as sheet metal for the rat guards was tested. Simultaneously, the village council discussed the issue of access by the poorest villagers to even the low-cost "package." It was finally decided to provide a village subsidy. The unspoken compromise that made this possible was the simultaneous design by the outside team of another system of storage improvements appropriate for wealthier villagers (storing up to five tons of grain as opposed to the average one ton required by poorer farmers).

Throughout the process of designing with villagers, the tension of working with a community actually composed of a range of socioeconomic levels and people involved with one another in patron–client relations was evident. For instance, during stage one it became evident that poorer farmers had less to store because they were forced by hunger to sell their labor to richer farmers at the beginning of the rainy season. Planting their own corn later as a result meant two things for the poorer farmers: (1) their corn yield was lower and (2) they had to harvest later, during a rainy month. This second fact meant that they harvested wet grain, subject to greater insect attack in storage. "Problem definition" sessions and "design" sessions did not follow up the logic of this clearly revealed vicious cycle in the village. Freire would call this a "limit situation" that should have become the explicit focus of debate and struggle. As it was, the politics of recent forced villagization precluded this, especially since the counterpart lay-expert team itself was almost totally composed of middle and influential peasants with only a minor representation of few poor peasants but no women.[1] As fruitful

---

[1]Although the outside team encouraged women to participate actively in all phases of the study, they were very reticent. The village council did not elect any women to the counterpart team, and while women attended the discussions, they seldom spoke up. The women were probably responsible for 90 percent of postharvest processing of grain in the village and were in frequent if not daily contact with storage structures because of

as this participation process was in terms of design product, it must be called merely instrumental in character. Poorer villagers were not empowered to challenge relations that locked them into their poverty.

CASE III: ELDERLY HOUSING AND A COMMUNITY CENTER FOR THE TEXAS FARM WORKERS' UNION

In 1981, the Texas Farm Workers' Union approached the School of Architecture and Urban Planning at the University of California Los Angeles (UCLA) for assistance in designing both housing for senior citizens and a community center. Because the senior citizens, retired from farm work, were living on extremely small incomes, their housing had to be buildable at very low cost:—$5,000 a unit. The area served by the Farm Workers Union is a strip of South Texas bordered by the Rio Grande and the Gulf of Mexico, close to the Republic of Mexico itself. The people there are primarily farm workers, most of Hispanic origin; many speak only Spanish, and the literacy rate is low. The summer climate is harsh, and high humidity prevents use of the cheapest building material, adobe.

Preliminary field reconnaissance in the area of Rangerville, Texas, and meetings with Union organizers indicated that a hands-on approach to participatory design was required in place of purely verbal "what do you want?" "what have you got?" exchanges. A modeling approach was decided upon, using extremely simple materials, with "doll house furniture" conveying a sense of scale. Union members of all ages were asked to come to an evening gathering in the Union hall. Because one advantage of the modeling approach is the opportunity to turn what might otherwise be a dull and unappealing event into a festive occasion, the gathering was presented not as an ordinary meeting, but as a party where, while the modeling was going on, food and drink would be available.

At the start of the gathering, participants in the modeling session were introduced to a "kit of loose parts" (Nicholson, 1974), consisting of cardstock, reusable adhesive, scissors, knives, pens, pencils, "Post-its," and other odds and ends. Previous experience had indicated that little initial explanation was necessary, as long as participants were clear on the "problem" to be "solved," and that participants enjoyed learning to

---

their cooking activities. The female member of the outside team, one of the present authors, made an effort to interview women privately. This yielded valuable insight but was, strictly speaking, outside of the participatory process. Local cultural norms, reinforced by the predominant Islamic faith, and the short duration of the study (three months) made it virtually impossible to attain the desired participation by women.

use the building materials in the first few minutes of a modeling session with the assistance of students or faculty. Participants were invited to work in groups of their own choosing, and did so. The "working sessions" were punctuated with a great deal of laughter, indicative of "success": people produce more ideas, and more innovative ideas, when they are "playing" with modeling materials than when they are "working hard."

The design products deviated markedly from the norm for senior citizen housing in American urban areas. The models generated by the UCLA students who had not been to Texas had one bedroom, while those constructed by the Texas farm workers had two. The traditional American norm envisions elderly people as a couple at most, or only one person; the Texas farm workers saw the elderly as an extended family. Furthermore, men often outlived women and remarried relatively young girls: thus, a 60-year-old man with a wife in her 20s and an infant at home was not unusual. Finally, in contrast to the American norm, the elderly members of the Texas farm workers were the most important members of the community. They were the center of activity, not just for their extended families but for the community as a whole. The Community Center was located by the farm workers adjacent to the elderly housing which, in turn, was in the center of the residential community. Such results might not have emerged from the surveys used in more traditional approaches to participation. We would, very simply, not have known what questions to ask.

Results were evident in process as well as product (Stea, 1981b). During the modeling process, a certain "leveling" took place in which experts were reduced to the same level as ordinary citizens; thus, a professional builder who expected others to follow his direction found himself to be just another participant. Women took the guiding role in several of the groups, even within this otherwise patriarchal society; wives who had always deferred to their husbands' judgment suddenly found themselves in positions of power in the layout of housing for their elderly relations. Because women were thus able to explore new power relations, one could say that this case lies intermediate between the Venezuelan and Tanzanian ones. There is more than instrumental use of process to achieve an appropriate low-cost product; more transformation is evident than in Bwakira Chini, but less than at Casalta II.

Case IV: Planning Primary Health Care in Western Kenya

The literature on participation in primary health care cited earlier is exemplified in the work of Miriam Were (Were, n.d.; Wisner, 1988). From a base at Nairobi University Medical School, she tested the limits of community participation in the design and implementation of community-based health care. This national pilot program ran between 1977 and 1980 in Kakamega District with the support of Kenya's Ministry of Health and UNICEF.

Kakamega was a difficult test. Land was very scarce. Average land-holding (1.4 hectare in 1979) was only half that which authorities calculated as the minimum necessary for subsistence. Partly as a result of such land poverty, infant mortality was high. The rate of outmigration of men seeking work was also very high. Remitted wages very nearly equalled the value of all agricultural production in the district and exceeded the combined value of sugar, maize, and beans in 1979. Roughly a third of households were headed by women.

Ninety-two community health committees were formed in as many small villages ranging in size from 70 to 376 households. These committees chose community health workers, who were trained and given drugs and dressings.

Central to the experiment was the local establishment of "community health funds" from which each community health worker was paid small stipends. The committees researched and discussed health problems in their communities, and a master list of health priorities was generated. The committees went beyond research, however, and a series of 52 concrete actions arose among the participating villages. These ranged from filling in potholes to eliminate mosquito breeding (malaria was ranked first in the overall inventory of problems) to the encouragement of breast-feeding.

As in the case of grain storage design discussed earlier, action research on health in Western Kenya faced, on occasion, limit situations, breaking through to what one might call second-order problems of poverty and power. For instance, faced with considerable land hunger, poor nutrition, and high child mortality, many of the local communities decided to focus energy on providing breakfast before school—a practice that was not common—and a school lunch program. This was linked to an impressive increase in the amount of intensive mixed vegetable cultivation near home and poultry production during the project period. Such innovations required little land. As in the Tanzania case, a fundamental contradiction blocking generalized rural development was avoided rather than confronted directly. Patterns of land ownership were not questioned, rather the local health research committee managed to find ways of increasing livelihood options of the land poor to some degree without addressing this fundamental inequality. However, as the majority of health committee members were women, and since they did create new economic and social "space" for themselves in the course of this project, the process was not without some transformative element.

CASE V: COMMUNITY LAND-USE MANAGEMENT IN ECUADOR

Sands (1987) describes an attempt to build on indigenous soil conservation practices. Although the problem—rates of erosion up to 20 times greater than the maximum acceptable as defined by U.S. Soil Conservation Service guidelines—was defined by outside agencies, implementation of this project was highly participatory. Many authors have emphasized the skill and local knowledge of small farmers (Altieri, 1987; Glaeser, 1984; Rau & Roche, 1988; Richards, 1985; Wisner, 1984, 1988). The Community Land Use Management Project (CLUMP) built carefully on existing cultural and agricultural practices that tended to conserve soil. For instance, patterns of fragmented land ownership dictated by inheritance customs—often seen as a problem for "agricultural modernization"—were accepted as actually working to conserve soil on very steep slopes because many small parcels were set apart, slowing the runoff of water.

In addition, between 1985 and 1987, local "conservation and credit" committees were established to screen requests for crop loans and manage the bank of improved seed provided by the project. These committees were composed of elected representatives of the communities and were also responsible for organizing local "field days" and training as well as for organizing communal work groups. The latter were based on the traditional reciprocal work groups in this Andean area known as *mingas*. Women extensionists were specially hired to liaise with female farmers, who are frequently excluded from such projects (Rogers, 1980). Details of implementation thus corresponded closely to the day-to-day needs of different gender and socioeconomic groups. No soil conservation measure was introduced that did not simultaneously increase income (Sands, 1987, p. 17).

The communities responded to local control and the class/gender sensitivity of extension with a high degree of support. All four kinds of participation identified by Shingi and his Indian co-workers in the context of social forestry were identifiable here (1986, pp. 133–152). There was "process participation" as local representatives were integrated into decision making. There was "cognitive participation" as the workings of local cultural and environmental systems were made clear to project staff by local people. There was "interactive participation" as local people were recruited as extensionists and communicators. There was "material participation" as communal work groups gave of their time and labor.

The complexity of land use, culture, economics, and politics in such intermontane valleys and highland plateaus would never have been incorporated into such a project without constant, active participation of the local people. In other cases, however, "indigenous knowledge" research seems only to have "packaged" what the outside agencies thought was correct and wanted in the first place. Some of the cases reported in the well-known collection by Brokensha *et al.* (1980) fall into this category, as does much of what calls itself "social marketing" of public health innovations (Manoff, 1985; Wisner, 1987).

The project relied on instrumental participation, to be sure, but there was also a strong element of local control, and serious determinants of poverty, such as reliance on remitted income by migrant men and youth, were directly discussed and confronted. Thus empowerment or transformation was also involved along side a more instrumental use of the participation process to achieve a sustainable soil erosion design.

IMPLICATIONS FOR FUTURE RESEARCH AND APPLICATIONS

This brief and partial review of action research and participatory methods should leave the reader with a sense of the very wide range of methods and techniques available as well as the great variety of situations in which they have been employed. It should also be clear that participation is not a panacea; hence, several research directions are clear.

First, this continuum of applications needs further documentation. In doing so, the communication between practitioners—often found in small development agencies, trade unions, farmer cooperatives, women's groups, and scholars—needs to improve.[2]

Second, one needs to clarify the ways in which instrumental and transformative methods of participation differ. In particular, how valid is

[2]Several attempts are underway, such as that by the Hesperian Foundation (Palo Alto, California), to circulate the grass-roots experience of participatory health projects. In the field of food security and women's empowerment, a similar attempt to bridge academic and grass-roots environments by the Food and Agricultural Organization's Freedom from Hunger Campaign (its *Ideas and Action* publication), the Luce Food Program at Hampshire College (its *African Food Security Trialogue*), and the Non-Governmental Organization Program of the International Institute for Environment and Development. The International Council for Adult Education (Toronto) maintains a network of participatory action researchers. For architectural design applications, the Environmental Design Research Association serves as a general clearing house; while in the realm of cross-cultural communication, the International Center for Built Environment (Santa Fe, New Mexico) is a reference point.

the sometimes expressed concern that transformative participation can leave a community vulnerable to governmental backlash (Pigozzi, 1982)? Is it true, on the other hand, that only transformative applications lead, in the long run, to development measured by increased autonomy of the everyday protagonists, increased social justice, and site-specific environmental sustainability? Or, as some have cautioned, can transformative applications sometimes only reflect changing fashions and rhetoric without actually empowering people to take more control of their lives (Gow & Vansant, 1983)?

A third major area of research concerns the limits of participatory action research. How important are economic and political conditions such as export prices or geopolitical interference, which may lie outside the reach of community members or even national officials? Long-term follow-up studies of the effects of the kinds of cases cited earlier would help to clarify such issues. Longitudinal study would also help to reveal whether participation is generalizable and whether people once involved in such action projects go on to become more active in other domains of their community life.

Fourth, systematic research is needed on the subject of communication. To what extent are the critiques of simple electronic and cybernetic models of communication valid? To what extent is communication affected by gaps separating researchers/designers from client/participants of different classes, ethnicities, gender, or age?

Finally, comparative study on all of the points above should be considered a high priority, especially including comparisons of attempts at participation in the USSR and Eastern Europe, Western Europe, and North America which might, in the age of *perestroika*, yield important insights.

REFERENCES

Abrahamsson, B. (1977). *Bureaucracy or participation: The logic of organization.* Beverly Hills, CA: Sage.

Altieri, M. (1987). *Agroecology.* Boulder, CO: Westview.

Aprovecho. (1980). *Helping people in poor countries develop fuel-saving cookstoves.* Eschborn, Germany: German Agency for Technical Cooperation.

Barber, B. (1984). *Strong democracy.* Berkeley: University of California Press.

Becker, F. (1977). *User participation, personalization, and environmental meaning: Three field studies.* Ithaca, NY: Cornell University Program in Urban and Regional Studies.

Bentz, B. (1988). Active user participation in the housing process. In D. Canter, M. Krampen, & D. Stea (Eds.), *Ethnoscape.* Aldershot, England: Gower.

Bhaduri, A., & Rahman, A. (Eds.). (1982). *Studies in rural participation.* New Delhi: Oxford/IBH Publishing.

Bhasin, K. (1978). *Breaking barriers: A South Asian experience in training for participatory development.* Report of the Freedom From Hunger/Action for Development Regional Change Agents Programme. Bangkok: United Nations Food and Agricultural Organization.

Blaikie, P., & Brookfield, H. (1987). *Land degradation and society.* London: Longman.

Blaut, J. (1967). Geography and the development of peasant agriculture. In S. Cohen (Ed.), *Problems and trends in American geography.* New York: Basic Books.

Bollinger, E. (1986). *CADD activities survey.* Houston, TX: University of Houston Association of Computer-Aided Design in Architecture.

Brandao, C. (Ed.). (1984). *Repensando a pesquisa participante.* Sao Paulo: Editora Brasiliense.

Brokensha, D., Warren, D., & Werner, O. (Eds.). (1980). *Indigenous knowledge systems and development.* Washington, DC: University Press of America.

Brown, G., & Novitski, B. (1986). Nurturing design intuition in energy software. In J. Turner (Ed.), *ACADIA Workshop '86 Proceedings.*

Campbell, D. (1987). Participation of a community in social science research: A case study from Kenyan Maasailand. *Human Organization, 46*(2), 160–167.

Cashdan, L., Fahle, B., Francis, M., Schwartz, S., & Stein, P. (1979). A critical framework for environmental change. In M. Francis (Ed.), *Participatory planning and neighborhood control.* New York: City University of New York, Center for Human Environments.

Chambers, R. (1983). *Rural development: Putting the last first.* London: Longman.

Chambers, R. (1985, December). *Normal professionalism, new paradigms and development.* Paper for the Seminar on Poverty, Development and Food: Towards the 21st Century, in honor of the 75th Birthday of Professor H. W. Singer, Brighton, England.

Churchman, A. (1987). Issues in resident participation. *Participation Network, 5*(Fall), 20–21.

Dandekar, H., & Feldt, A. (1984). Simulation/gaming in Third World development planning. *Simulation and Games, 15*(3), 297–304.

Davidoff, P. (1965). Advocacy and pluralism in planning. *Journal of the American Institute of Planners, 31*(November), 331–338.

Derrida, J. (1978). *Writing and difference* (A. Bass, Trans.). Chicago: University of Chicago Press.

Dubell, F., Erasmie, T., & De Vries, J. (Eds.). (1981). *Research for the people/research by the people.* Linköping, Sweden: Linköping University, Department of Education.

Feldman, R. (1984). Women's groups and women's subordination: An analysis of policies towards rural women in Kenya. *Review of African Political Economy, 27/28,* 67–85.

Field, A. (1986). *The Cayuga Indian land claim: Application of a community-based natural resource inventory process.* Unpublished masters thesis, Cornell University, Ithaca, NY.

Foucault, M. (1980). *Power/knowledge* (C. Gordon, Ed.). New York: Beacon.

Francis, M. (1979). *Participatory planning and neighborhood control.* New York: City University of New York, Center for Human Environments.

Freire, P. (1973). *Pedagogy of the oppressed.* New York: Seabury.

Gandhi, M. (1952). *Rebuilding our villages.* Ahmedabad, India: Navajivan Publishing House.

Gardner, N. (1974). Action training and research: Something old and something new. *Public Administration Review, 34,* 106–115.

Glaeser, B. (1984). *Ecodevelopment in Tanzania.* Berlin: Mouton.

Gow, D., & Vansant, J. (1983). Beyond the rhetoric of development participation: How can it be done? *World Development, 11*(5), 427–446.

Gran, G. (1983). *Development by people: Citizen construction of a just world.* New York: Praeger.

Guggenheim, H., & Fanale, R. (1975). Water storage through shared technology: Four projects among the Dogon in Mali. *Assignment Children, 45/46,* 151–166.

Habermas, J. (1975). *Legitimation crisis* (T. McCarthy, Trans.). Boston: Beacon Press.

Hall, B., Gillette, A., & Tandon, R. (Eds.). (1982). *Creating knowledge: A monopoly?* New Delhi: Society for Participatory Research in Asia.

Herrera, A. (1984). *Project on research and development systems in rural settings: Final report.* Tokyo: United Nations University.

Hilton, D. (1983). 'Tell us a story': Health teaching in Nigeria. In D. Morley, J. Rohde, & G. Williams (Eds.), *Practicing health for all* (pp. 145–153). Oxford: Oxford University Press.

Himmelstrand, U. (Ed.). (1981). *Spontaneity and planning in social Development.* Beverly Hills, CA: Sage.

Huizer, G. (1984). Harmony vs. confrontation. *Development: Seeds of Change, 2,* 14–17.

Jacome, S. (1988). Environmental modelling: The view from Ecuador. In D. Canter, M. Krampen, & D. Stea (Eds.), *Ethnoscape.* Aldershot, England: Gower.

Kalyalya, D., Mhlanga, K., Semboja, J., & Seidman, A. (1988). *Does aid work? A participatory learning process in Southern Africa.* Trenton, NJ: Africa World Press/Oxfam-America.

Kantrowitz, M., & Nordhaus, R. (1980). The impact of post-occupancy evaluation research: A case study. *Environment and Behavior, 12*(4), 508–519.

Kaplan, R. (1987). Simulation models and participation: Designers and clients. In J. Harvey & D. Henning (Eds.), *Public environments.* Washington, DC: Environmental Design Research Association.

Kaplan, S. (1977). Participation in the design process: A cognitive approach. In D. Stokols (Ed.), *Perspectives on environment and behavior.* New York: Plenum.

Kruks, S. (1983). *Notes on the concept and practice of 'participation' in the KWDP (with special emphasis on rural women).* (Kenya Woodfuel Development Project Discussion Paper.) Nairobi and Stockholm: The Beijer Institute.

Kruks, S., Rapp, R., & Young, M. (Eds.). (1989). *Promissory notes: Women and the transition to Socialism.* New York: Monthly Review.

Kruks, S., & Wisner, B. (1984). The state, the party, and the female peasantry in Mozambique. *Journal of Southern African Studies, 11*(1), 106–127.

Lamming, G. (1980). *Women in agricultural cooperatives: Constraints and limitations to full participation.* Rome: United Nations Food and Agricultural Organization.

Lawrence, R. (1983). Laypeople as architectural designers. *Leonardo, 16*(3), 232–237.

Lawrence, R. (1988). Environmental modelling for house planning. In D. Canter, M. Krampen, & D. Stea (Eds.), *Ethnoscape.* Aldershot, England: Gower.

Lewin, K. (1946). Action research and minority problems. *Journal of Social Issues, 1–2,* 34–36.

Longhurst, L. (Ed.). (1981, October). Rapid rural appraisal. *IDS Bulletin, 12*(4).

McKechnie, G. (1977). Simulation techniques in environmental psychology. In D. Stokols (Ed.), *Perspectives on environment and behavior.* New York: Plenum.

Manandhar, R. (1986). *The role of self-reliance in small communities.* Unpublished doctoral dissertation, University of Melbourne.

Manoff, R. (1985). *Social marketing: New imperative for public health.* New York: Praeger.

Mao Tse-tung. (1967). The united front in cultural work. *Selected works of Mao Tse-tung.* Beijing: Foreign Languages Press.

Mansbridge, J. (1980). *Beyond adversary democracy.* New York: Basic Books.

Marx, K., & Engels, F. (1970). *The German ideology* (C. Arthur, Ed.). New York: International Publishers.

Maxwell, S. (1984). Farming systems research: Hitting a moving target. *IDS Sussex Discussion Paper 199.* Falmer, Brighton: University of Sussex.

Mbilinyi, M., Vuerla, U., Kassam, Y., & Masisi, Y. (1982). The politics of research methodology in the social sciences. In Y. Kassam & K. Mustafa (Eds.), *Participatory research:*

An emerging alternative methodology in social science (pp. 34–63). New Delhi: Society for Participatory Research in Asia.

Mduma, E. (1982). Appropriate technology for grain storage at Bwakira Chini Village. In Y. Kassam & K. Mustafa (Eds.), *Participatory research: An emerging alternative methodology* (pp. 198–213). New Delhi: Society for Participatory Research in Asia.

Morley, D., Rohde, J., & Williams, G. (Eds.). (1983). *Practicing health for all.* Oxford: Oxford University Press.

Mwansa, D., & Kidd, R. (Eds.). (1982). *Third World Popular Theatre Newsletter, 1*(1) [c/o Dickson Mwansa, DEMS/UNZA, P. O. Box 20350, Kitwe, Zambia].

Negroponte, N. (1970). *The architecture machine: Toward a more human environment.* Cambridge, MA: MIT Press.

Negroponte, N. (1975). *Soft architecture machines.* Cambridge MA: MIT Press.

Newell, K. (Ed.). (1975). *Health by the people.* Geneva: World Health Organization.

Nicholson, S. (1974). The theory of loose parts. In G. Coates (Ed.), *Alternative learning environments: Emerging trends in environmental design.* New York: Plenum.

Nyerere, J. (1968). Education for self-reliance. In J. Nyerere, *Freedom and socialism/Uhuru na ujamaa* (pp. 267–290). Dar es Salaam: Oxford University Press.

Palmer, E. (1985). *The environmental intervention process: A cross-cultural approach to architecture and development.* Unpublished masters thesis, University of Wisconsin-Milwaukee.

Pateman, C. (1964). *Participation and democratic theory.* Cambridge, England: Cambridge University Press.

Patricios, N. (1979). An agentive perspective of urban planning. *Town Planning Review, 50*(1), 35–54.

Peattie, L. (1968). Reflections on advocacy planning. *Journal of the American Institute of Planners, 34,* 80–88.

Peattie, L. (1970). Community drama and advocacy planning. *Journal of the American Institute of Planners, 36,* 405–410.

Pigozzi, M. (1982). Participation in non-formal education projects: Some possible negative consequences. *Convergence, 15*(3), 6–18.

Pyatok, M., & Weber, H. (1978). Participation in residential design. In H. Sanoff (Ed.), *Designing with community participation.* New York: Plenum.

Rahman, A. (Ed.). (1984). *Grass-roots participation and self reliance.* New Delhi: Oxford/IBH Publishing.

Rand, G., Heath, P., & Wilde, M. (1977). Research agenda: Coming to terms with the environment's future. In S. Weidemann & J. Anderson (Eds.), *Proceedings of the Environmental Design Research Association.* Washington, DC: Environmental Design Research Association.

Rau, B., & Roche, S. (1988). *Working for the food of freedom: African initiatives for change.* Washington, DC: Africa Faith and Justice Network.

Richards, P. (1985). *Indigenous agricultural revolution.* London: Hutchinson Education.

Rogers, B. (1980). *The domestication of woman.* London: Tavistock.

Sanchez, E., Cronick, K., & Wiesenveld, E. (1988). Psychological variables in participation: A case study. In D. Canter, M. Krampen, & D. Stea (Eds.), *Ethnoscape.* Aldershot, England: Gower.

Sands, M. (1987). Hillside soil conservation on farms in Ecuador. In K. Tull (Ed.), *Experiences in success.* Emmaus, PA: Rodale International.

Sanoff, H. (1979). *Design games.* Los Altos, CA: Kaufmann.

Schaffer, B., & Lamb, G. (1981). *Can equity be organized? Equity, development analysis and planning.* London: Gower.

Shannon, C., & Weaver, W. (1949). *The mathematical theory of communication.* Urbana: University of Illinois Press.

Shingi, P., Patel, M., & Wadwalkar, S. (1986). *Development of social forestry in India.* New Delhi: Oxford/IBH Publishing.

Simon, J. (1988). Participation and community: Transcultural perspectives. In D. Canter, M. Krampen, & D. Stea (Eds.), *Ethnoscape.* Aldershot, England: Gower.

Smith, R. (1978). *Public participation in planning and design: Implications from theory and practice for the design of participatory processes.* Unpublished doctoral dissertation, University of California, Berkeley, CA.

Sommer, J. (1977). Action research. In D. Stokols (Ed.), *Perspectives on environment and behavior.* New York: Plenum.

Stea, D. (1980). Environmental modelling as participatory planning. *Fourth World studies in planning, 5.* Los Angeles: University of California at Los Angeles, School of Architecture and Urban Planning.

Stea, D. (1981a). Human energy and participatory design. *Energy Resource Journal, 1*(2), 26–27.

Stea, D. (1981b). Participatory planning and design of Waahi Marae. In N. Ericksen (Ed.), *Environmental perception of planning in New Zealand.* Hamilton, New Zealand: University of Waikato.

Stea, D. (1982). Cross-cultural environment modelling. In A. Lutkus & J. Baird (Eds.), *Mind, child, and architecture.* Hanover, NH: University Press of New England.

Stea, D. (1984). Participatory planning and design for the Third World. In W. Gilland & D. Woodcock (Eds.), *Architectural values and world issues.* Silver Spring, MD: International Dynamics.

Stea, D. (1988). Participatory planning and design in intercultural and international practice. In D. Canter, M. Krampen, & D. Stea (Eds.), *Ethnoscape.* Aldershot, England: Gower.

Swantz, M.-L. (1975). Research as an educational tool for development. *Convergence, 7*(2), 44–53.

Thompson, J. (1984). *Studies in the theory of ideology.* Cambridge: Polity.

Turner, J. (1984). *The architect as enabler of user house planning and design.* Stuttgart, West Germany: Karl Dramer Verlag.

UNESCO. (1981). *Domination or sharing? Endogenous development and the transfer of knowledge* (Series: Insights). Paris: UNESCO Press.

Vio Grossi, F. (1981). Socio-political implications of participatory research. *Convergence, 14*(3), 43–51.

Vio Grossi, F., Gianotten, V., & de Wit, T. (Eds.). (1981). *Investigación participativa y praxis rural: Nuevos conceptos en educación y desarrollo comunal.* Lima: Mosca Azul Editores.

Wandersman, A. (1987). Research on citizen participation. *Participation Network, 5*(Fall), 22–25.

Weisman, G. (1983). Environmental programming and action research. *Environment and Behavior, 15,* 381–408.

Were, M. (n.d.). *Organization and management of community-based health care.* Report of a National Pilot Project of Kenya Ministry of Health/UNICEF. Nairobi: UNICEF.

Wisner, B. (1970). Advocacy and geography: The case of Boston's urban planning aid. *Antipode, 2,* 25–29.

Wisner, B. (1984). Eco-development and eco-farming in Mozambique. In B. Glaeser (Ed.), *Eco-development: Concepts, projects, strategies* (pp. 157–168). London: Pergamon Press.

Wisner, B. (1987). Doubts about 'social marketing.' *Health Policy and Planning, 2*(2), 178–179.

Wisner, B. (1988). *Power and need in Africa: Basic human needs and development policies.* Trenton, NJ: Earthscan Publications and Africa World Press.

Wisner, B., Neigus, D., Mduma, E., Kaisi, T., Franco, L., & Kruks, S. (1979). Designing storage systems with villagers. *African Environment, 3*(3/4), 85–95.

9

Design Research in
the Swamp
TOWARD A NEW PARADIGM

JAY FARBSTEIN and MIN KANTROWITZ

> In the topography of practice, they [researchers] must either remain on the
> hard dry ground where they can function according to the canons of tech-
> nical rationality, applying research-based theory and technique to a narrow
> range of well formed problems . . . of problems of limited scope and impor-
> tance . . . or they can descend to the swampy terrain below where they
> cannot be rigorous in any sense they know how to describe. (Donald Schon,
> 1984, p. 4).

The chapters in this volume treat advances in environment, behavior,
and design. But what are "advances" in *applied* environmental design
research? Discussing this question, we realized that its very formulation
implicitly assumes both that research is a separate activity from design
and that research precedes design. Under this paradigm, the researcher
operates as an academic or a consultant, submits a report or writes a
paper, and then, most likely, disappears from the scene while someone
else reads, understands, and finally applies the results.

This model reflects a traditional epistemology of design research, in
which research-based knowledge is applied to a design problem, much
like a band-aid is applied to a bruise. Band-aids are useful and impor-

---

Jay Farbstein • Jay Farbstein & Associates, Inc., 1411 Marsh Street, Suite 204, San Luis
Obispo, California 93401. **Min Kantrowitz** • Min Kantrowitz & Associates, Inc., P.O. Box
792, Albuquerque, New Mexico 87103.

tant—but not sufficient—and this paradigm alone does not adequately describe either recent developments in design research practice or its future potential. The most recent advances we perceive in design research represent not only movement forward in an established direction but also what seems to us to be the beginning of a redefinition of the field.

What then are these advances in applied environmental design research? Where is applied design research practice today? We see it as happening in that "swampy terrain," described in the quotation above by Donald Schon, where pressure for answers is constant and professional boundaries are often unclear. In the swamp, roles of individuals and groups are growing and changing, issues swim by with amazing rapidity and complexity, and the air is thick with implications. Methodological rigor may be rarely achieved, but intellectual rigor, problem-solving ability, and communications skills are paramount. Models, or even patterns, which describe this practice of design research may be difficult to identify, but that is what we are setting out to do.

Thus, this chapter is an exploration and discussion of a new paradigm for design research rather than a description of design research within a traditional applied research model. We realize that, even as practiced according to its traditional model, design research still is far from fully integrated into standard design practice. We hope that as this chapter unfolds it will become clear—at least by contrast—that part of the reason that traditional design research has had limited acceptance is due to its nature as essentially divorced (in time and persona) from the arena of design decision making.

We wish to introduce the concept of what we call "design-decision research" and to answer basic questions about it: What is it? Who does it? When in the building life cycle does it occur? How is it being performed? and Why is design research evolving in this direction? We give examples based on our knowledge of design-decision research in North America today and begin to explore the implications of this evolving paradigm. (Where examples have been published, references are cited; however, many of our examples, by their nature, are unpublished.)

DESIGN-DECISION RESEARCH

Design is a process of creating, exploring, and testing physical options to accomplish a given set of objectives, based on the designer's experience, skill, and knowledge. Research is (and should be) only one source of this knowledge. For each decision, for example, exploring site plan issues or specifying floor tile, the designer goes through an iterative process, first identifying options, then examining their implications, successively narrowing choices until the decision is made. In the applied design research model, however, the researcher stands outside the decision framework. Research results are produced that either comment on the outcomes of the designer's choices (as in a POE report), attempt to influence the choices before they are made (as in programming), or address a conceptual environment and behavior issue (such as wayfinding) which may have generic, though perhaps not immediate, implications for design choices.

For the two former types of applied research, information would be developed in one of two ways: either existing research is consulted (for input, support, affirmation), or new research is commissioned (to gather information relevant to the specific project). Research consultants perform the analysis and present the results. Then the results are applied by the designer at the appropriate point in the design cycle, generally as a one-time-only intervention, and the design process continues. We call this a "consultative" model, largely because the consultant is so often removed from the arena of design decisions (not because the consultant is an independent professional). The implied relationship between research and design is illustrated most thoughtfully in Zeisel (1975), who was perceptive enough to see the process as cyclical and iterative (Figure 1).

The consultative approach is still being widely practiced and advocated. It underlies much of the current literature on postoccupancy evaluation (Preiser, Rabinowitz & White, 1988). The consultative approach maintains independence and clarity; it is practiced on Schon's "hard, dry ground . . . of well formed problems." But, in an increasingly complex, competitive environment, with growing pressure for profitable and timely design excellence, this clarity is becoming characteristic of an ideal situation, compared to the reality of the swamp.

On hard dry ground, job responsibilities are clearly defined. For example, according to the standard American Institute of Architects (AIA) form of agreement between architect and owner (AIA, 1980), the architect offers five basic services: schematic design, design development, construction documents, bidding/negotiation, and construction contract administration. Yet, many architects are involved with much more than this, offering "supplementary" or additional services such as project feasibility, programming, construction management, and a host of others. Architects' responsibilities are being remodeled as the profession undergoes drastic change. As Robert Gutman (1988) points out, many of these changes are in response to such factors as the rising number of architects (competition spawning a need to specialize in cer-

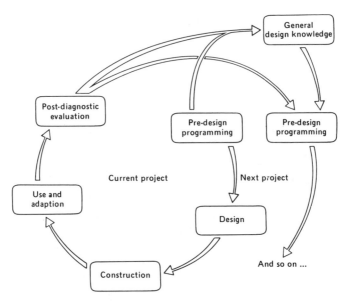

FIGURE 1. Five steps in the design cycle. Taken from *Sociology and Architectural Design* by John Zeisel. © Russell Sage Foundation, 1975. Used with permission of the Russell Sage Foundation and the author.

tain services or building types), the increasing size of firms (which allows and promotes specialization), the organizational characteristics of clients (who demand more services), and the increasing complexity of buildings.

The boundaries delineating the activities of design practice, research practice, and the practice of other associated disciplines are becoming increasingly muddy. The roles of designer, engineer, planner, technical consultant, legal advisor, researcher, facility manager, contractor, and financier are starting, under certain conditions, to overlap. Thus, research-based knowledge is being used (or can be used) in all phases of a project and by people with many different job titles. This seems to us to be something new, and we have chosen to call the polymorphous research that is responsive to current conditions "design-decision research" to distinguish it from research which may be of equal (or perhaps greater) value, but which is not developed in as demanding a decision-making process. Rather than to develop new knowledge, the goal of design-decision research is to better inform the decision-maker. (A recent special issue of the *Journal of Architectural and Planning Research* called one aspect of this research "demand-driven post-occupancy evaluation" [Zimring, Wineman, & Carpman, 1988].)

How is Design-Decision Research Different from Action Research?

To launch our definition of design-decision research, we wish to distinguish it from action research, which it slightly resembles. In the early years of design research practice, when social change was explicitly acknowledged as a goal, some researchers embraced the 40-year-old Lewinian action research tradition. In a recent book, reviewing applications of Lewin's Field Theory in current practice, *action research* is defined as "applied research which integrates together scientific methodology with the consultative process" (Woods & Casper, 1986).

According to Kurt Lewin, founder of action research, knowledge is advanced and problems are solved through close cooperation between the researcher and the practitioner. "Practitioners keep theorists in contact with social reality and theorists provide practitioners with a deeper understanding of the social problems that confront them" (Johnson & Johnson, 1982, p. 14). This cooperation, in an action research model, "is based on the assumption that social change efforts are enhanced through systematic data collection around some perceived problem or goal" (Stivers & Whellan, 1986, p. xii). Thus, in an action research model, one can consider design as the physical manifestation of social goals (a type of social action through design). While this model was a radical departure from the "isolated researcher" approach prevalent prior to Lewin's work, it still assumes that research is not integral to change, but is separate from it in time and person and "enhances" it. The developing practice of design-decision research eliminates that separation.

Design-decision research is a different way of thinking and acting in relation to the design process. In this model, "design" is defined broadly as a multidimensional, information-based activity that may start with the initial feasibility study for a project and continue through occupancy, possibly throughout the building's entire life cycle. Design-decision research can be an integral part of any decision relating to the facility and its use. This model recognizes that many factors influence design. There are environment–behavior concerns (such as user needs and environmental psychological factors) and architectural concerns about aesthetic, functional, and technical issues. In addition, there are economic, administrative, political, and managerial concerns that are closely inter-

woven with the built and occupied environment. Design-decision research expands the focus beyond the five design steps defined by the
AIA into a broader range of concerns including financial feasibility, legislative action, policy development, asset management, and regulatory
enforcement. In this model, the entire *context* of design can be the subject of design research, expanding the potential for the influence of
design-decision research from a set of defined contact points to a continuum. Research and design activities may be performed simultaneously rather than sequentially, perhaps by the same individual. Figure 2 attempts to illustrate the design-decision research model (though
its full complexity probably cannot be indicated in two dimensions).

The design-decision research paradigm can best be understood by
looking at the answers to five questions. These are briefly stated here,
and discussed next in more detail.

- What is it? It is research focused on design decision making.
- Who does it? Design-decision research may be contributed to or
  carried out by people from a wide variety of professional backgrounds; job titles are not good predictors of project activities or
  responsibility.
- When does it occur? Design-decision research can occur at any
  time (or many times) in a project; it does not necessarily precede
  design.
- How are results presented? Since good communications facilitate
  good decisions; results are presented in a variety of ways (not
  necessarily as research reports).
- Why is design research evolving in this direction? Through diffusion of innovations, organizations are finding that design research
  helps decision-makers.

What Is Design-Decision Research?

Design-decision research is, by definition, research which is consciously directed toward contributing to design decisions. It focuses
explicitly on helping clients realize their objectives. Rather than approaching an issue by analyzing all its components, design-decision
research asks: What are the critical issues here? What decisions will be
made based on information to be developed? The activities of the researcher depend directly on the answer to these questions, rather than
on a predetermined agenda or approach (such as the researcher's interest in a theoretical issue or methodological approach).

The researcher's role is to help the organization makes *its* own best
decisions, within the context of *its* objectives. In a litigious and rapidly

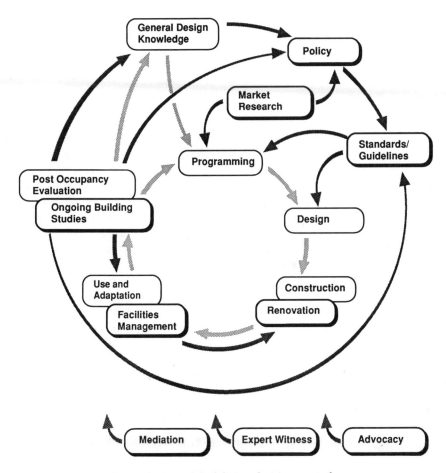

FIGURE 2. A model of design-decision research.

changing financial, legislative, and regulatory environment, this approach appeals to decision-makers precisely because it is practical and businesslike. It results in answers that help people make decisions; it does not necessarily result in studies or reports. Design-decision researchers strive "to supply answers to the day-to-day questions facing managers and administrators of building management and construction programs" (Zimring *et al.*, 1988, p. 274). As a result, research-based design decision making is being more widely used within organizations.

Design-decision research provides a key part of the kit of tools needed for a systematic approach to facilities management and long-range facilities planning. Since design-decision research can develop information germane to any phase of a building's life cycle, it can contribute to strategic planning, programming, feasibility studies, design, construction, operation, fine tuning, renovation, maintenance, repair, and so forth (Farbstein *et al.*, 1986).

Design-decision research helps make the client's concerns and goals explicit. It is one way for an organization to question and clarify its values; for example: What image does it want? How can it resolve trade-offs between customer and employee needs? Should it pay more attention to first costs or investment for the future (Moleski & Lang, 1982)? Design-decision researchers trained in environment–behavior methods use them to help organizations confront these questions directly. For example, a bank's top management decided to offer a new set of services to their most valued customers (those with a high net worth and need for capital). To support this effort, management was about to undertake major renovation of one wing of the bank. They hired an environment–behavior researcher to help them decide how to design the new "high-value" customer area to help those customers feel comfortable and to reflect the high status management felt the area should project. Thus, this intervention was set up according to a traditional design research model. However, through in-depth interviews and focus groups, the underlying needs of the management (to keep those valued customers) and customers (to feel that they were treated personally as valued individuals) were identified—and found to be in conflict with the originally intended renovation. Based upon the new information, the bank's design goals changed from expressing a high-status image to providing places for individual, personal contact with banking representatives. The ability to uncover, recognize, and precipitate this shift in direction is the essence of design-decision research.

Although some of the features of this approach have been described elsewhere, particularly the focus on recognizing organizational needs and goals in research (Zeisel, 1986), prior approaches differ from design-decision research in that they tend to accept the goals as stated and carry out design research to retrospectively test how well those goals were reached. In other words, client goals are recognized, but the client is not invited to question them.

Since it consciously strives to contribute to clients' objectives, design-decision research inevitably confronts issues of organizational structure, power, and politics that can suggest directions for research management. There is always a question of the potential political use (and volatility) of research findings. If not addressed, issues such as who wants the research done and why, or who may benefit or suffer from results, can render the research useless. Therefore, in any organization with different interest groups (each of which may have its own agenda and reward system), it is essential for the design-decision researcher to address questions like: Whose interests are being served by the research? Who will have control over findings? What is the relationship between managerial "sponsorship" of research and those with the power to implement findings and recommendations? (These issues are discussed in greater depth in Farbstein, Kantrowitz, Schermer, & Hughes-Caley, 1990.)

With a decision-making focus, design-decision research is occurring in a wide variety of contexts. Examples include researchers working in settings such as the following:

Public administration. A top-level administrator in a state agency sets housing policy based upon existing research findings and commissions new studies to answer policy questions. With a focus on building quickly and economically while still meeting occupant needs as identified by extensive research, there is constant pressure to make decisions about the relative priority of competing objectives.

Product design. A design researcher determined needs for, designed, analyzed, and tested new furniture systems for well and infirm elderly users. Given a goal of enabling greater independence and activity levels, design-decision research provided input into each phase of the development process for the "geriatric personal furnishings system," a complete system of furnishings and equipment designed to maximize self-rehabilitative behaviors (Koncelik, 1982; NEA, 1984).

Retail business. A design-decision researcher evaluates the marketing impacts of design decisions by examining the relationship between customer demographics and the design of hospitality environments (hotels and restaurants). As a result of the research, certain building exterior and interior designs are used in some markets, but not in others.

Facility management and real estate. An office leasing and management firm evaluated indicators of satisfaction with office environments on a regular basis as a way to maintain competitive position relative to other office leasing organizations. They offered these services to prospective tenants as part of the process of determining requirements for tenant improvement packages (G. Robertson, personal communication, April 1988).

WHO DOES DESIGN-DECISION RESEARCH?

Not long ago, conventional wisdom held that designers were responsible for design and researchers for research, prompting a raft of academic discussions about the need for translation, about the "gap" and the "squishy middle." To some extent, this gap has been bridged by the emergence of a "new practitioner" (see Schneekloth, 1987), who might have dual training in design and a social science or, even more promisingly, might have been trained in design-decision research itself. Recently, a writer has described the latter as the "one community" model versus the former "two community" model of separate discipline areas (Min, 1988). We would argue that, simply by knowing a participant's professional background or role, it is difficult to predict who will take responsibility for advocating design research concerns.

Design projects today involve an increasingly large and diverse group of people, each with their own unique education, professional mind-set, experience, and expectations about their role in the project. Project coordination and information management are becoming roles in themselves, at the same time as there are increasing conflicts about responsibilities, liability, and turf among traditionally defined disciplines (Gutman, 1988; Joroff & Moore, 1984). Further complicating all of this is the "*Rashomon*" phenomenon, wherein various parties each have a different view of the ultimate objective, their role, and the process they will experience together (and perhaps, when it's over, of what actually happened). Under financial and time pressures, the swamp can become even muddier.

What is the role of the design-decision research practitioner in these situations? There is no single—or simple—answer, since the design-decision approach considers that decisions about a wide variety of issues could be facilitated or informed by research knowledge. Thus, design-decision research practitioners now work under a wide variety of job titles, scopes of work, and project responsibilities (and may even change roles on different projects). In fact, the design-decision research practitioner may not be an outside professional consultant, but may be an employee, manager, or member of the community. Here are some examples of roles which may be taken and activities which may be carried out.

Facility programmer. The programmer is on the "front line," working intensively with clients and users to develop behaviorally based information as input to design. During the process of defining the project, however, concerns from every other phase and discipline can be considered (and may be critical). For example, in programming a hospital, technical and user issues needed to be considered simultaneously in order that their intersecting requirements were articulated (Carpman, Grant, & Simmons, 1986).

Architect. While the architect has primary responsibility for design and coordination of a team of engineers and other specialists, in the design-decision research model, the architect may help define issues that need exploring, options that need testing, or limitations that must be respected. The architect will also be the one who assures that research-based information is integrated into the final design.

Market analyst. The researcher assists marketing staff to assure that facility design reflects the organization's marketing approach (which derives from its superordinate goals). The United States Postal Service followed this technique in developing guidelines for future postal lobbies. These were to emphasize self-service and clear communication of marketing and public-relations messages (United States Postal Service, 1987).

Communications consultant. The consultant works with media personnel to develop effective tools to communicate information to decision-makers in the organization and to the public. In settings where wayfinding is both difficult and crucial, design-decision researchers have been part of the team with communications experts to design environmental, management, and graphics systems to support wayfinding (Carpman, Grant, & Simmons, 1984).

Facility manager. As an in-house resource, the facility manager may take responsibility for examining patterns of facility use and interpreting their implications for future facility needs. The facility manager may be the channel of communications from users in the organization to upper management.

Mediator. A mediator may help organizations resolve differences, before they become involved in adversarial proceedings, by facilitating design decision making among conflicting user groups. In one example, multiple agencies forced to share a new building with extremely limited space found—through the help of a mediator—ways to share administrative support rather than battle over who got what space (Moreno, 1989).

Community advocate. The researcher develops information that may help to define the scope and nature of the problems a community or interest group faces—then helps the community use the findings as a lever in the political arena (Wandersman, 1980).

Expert witness. The researcher may be called upon to testify in court as to the environment and behavior effects of an antiquated city jail. In the legal arena, standards of proof are different than elsewhere, and competing experts may be called to question the strength and solidity of findings and their implications for this particular jail and its inmates. Based upon the results of the testimony, the jail is closed and the city is forced to spend millions of dollars to rebuild it according to a set of principles defined in part by the research. Another example of the impact of the researcher on decisions, taken from the housing field, is reported by Bechtel (1978):

> The essence of the suit against the developer was that he had failed to present design alternatives to the high-rise in his environmental impact statement. Clare [Cooper-Marcus] was asked to testify against the high-rise and to present design alternatives. . . . She realized in her testimony that the lawyers could not deal with the research evidence she presented so they tried to discredit her. . . . The court decision ruled against the developer, and the 70-page opinion of the court frequently quoted from Clare's testimony. The case was precedent-setting in that it established the developer must be accountable for the human consequences of his development. (p. 438)

The situation gets even more interesting when a project team, assembled with people from backgrounds like these (and others), sits down in a room together and is confronted with the need to complete a project. Perhaps this is the reason that, increasingly, design researchers help organizations manage the *process* of facility design and operation, rather than simply providing information about a behavioral issue or building type. These changing roles require more than knowledge; they require the ability to analyze and understand the design situation from many points of view and to intervene as a facilitator with the kind of action or information that is needed at a given moment. As the next section discusses, that moment may occur throughout the design cycle.

WHEN DOES DESIGN-DECISION RESEARCH OCCUR?

In the design-decision research model, the research work can occur at any time (or many times) during a project; it no longer simply precedes or follows design. This kind of research is stimulated by demand that can potentially occur at any point in the process of planning, development, design, construction, occupancy, and management of sites and facilities. Of course, the kind of information or intervention needed can vary depending on the phase. Thus, design-decision research information must be ready for delivery close to the time when it is needed. This is in marked contrast to "doing a study" of an issue, a building, or a building type and presenting results to someone else for them to implement (or for their edification). When the value of this type of input is recognized and built into the organization, it amounts to institutionalization of an information-based design and management approach, using research in many phases of design decision-making where it had not been done before.

The following examples illustrate new areas of involvement—project phases in which design-decision research practitioners are now active.

Project financing review. The researcher consults to a bank, advising on whether to give loans to developers of specific projects. Design proposals are evaluated on the basis of their utilization of (or performance in terms of) research-based design criteria. The bank is learning that these factors are key to the projects' acceptance and, therefore, financial success.

Value engineering. The researcher works, in effect, as a mediator, facilitating value engineering decisions between owners, architects, and engineers for a state corrections facility. Value engineering (as opposed to cost engineering) seeks not simply to reduce first costs of construction but to evaluate life cycle costs for a facility and to find lower cost options that support all necessary functions of the facility. Having a researcher in this role broadens the definition of "value" beyond minimum levels of building function to include psychological variables and programmatic objectives. For example, a snack bar that had been considered for elimination based solely on its cost was found to be valuable enough to retain because of its implications for quality of life for inmates.

Corporate merger. The researcher helps one party of a potential corporate merger to identify its future facility-related goals, in order to enter merger negotiations from a position of clarity and strength. With an aging facility and limited capital, the merger became an attractive way of solving facility problems.

Litigation. The researcher was called as an expert witness in cases covering accidents on stairs, housing project residents' security, inmates' conditions of confinement, relocation of the elderly, and citizens' protection from environmental hazards (Pauls, 1986).

> One of the most dramatic cases and one with major impact involved the development of a preparation program to reduce the consequences of involuntary relocation for institutionalized elderly persons. Research by Dr. Norman Bourestom . . . and Dr. Leon Pastalan . . . documented that the mortality rate doubled for elderly persons who were involuntarily transferred from one institutional setting to another. . . . Accordingly, the Michigan group developed a program to prepare persons for involuntary relocation that would reduce the fatal consequences of transfer trauma. . . . The Michigan findings and recommendations . . . served as the basis for court decisions in seven civil rights cases in five states. (Archea & Margulis, 1979, p. 218)

Architectural competition/prize. The researcher prepared the program, recommended an evaluation process, advised the jury, and assessed submissions (Jockusch, in press). A similar role was played by researchers in helping define objectives and review submissions for an award in urban excellence (the Bruner Prize).

How Are Design-Decision Research Results Presented?

Since design-decision research is geared toward decision making, effective communication of results is essential. Good communications facilitate good decisions. Effective communication begins with understanding what decisions will be made, and how the research might influence those decisions. By analyzing the audience for the communication—its needs, ambitions, biases, education level, political orientation, and so forth—the communication can be designed to optimize its impact.

More emphasis is being placed on two-way communication. It is less often the role of the consultant to have all the answers, to be the expert. More often, the most important thing the consultant can do is ask the right questions, or facilitate a process of group decision making and concensus building.

When research findings are presented, briefer, more focused communication methods are being used. Often, results are delivered in person, in a short, punchy "briefing" to decision-makers, or in an outline style executive summary that begins with conclusions and recommendations, relegating details of research design and even findings to appendices. The goal is to tailor the communication to the audience; to clearly show information, research, and design in a way that will contribute to making decisions. Audio, video, and electronic media are increasingly being used to heighten the impact of the message. For design-decision research, the "report" is being redefined, from a formal noun (the probably dusty, bound volume) to an informal verb ("to tell").

Privately funded design-decision research results may never appear in public because they are considered proprietary information. In other situations, research results may never be published at all since there may not be a report; rather, design changes may be made directly on the drawings or even in the building, based on findings and recommendations. The emphasis for designing the communication is on the *action* that may occur as a result of the research, not on the research itself. (Of course, this does not obviate the need to document the decision process, a practice that is apparently unavoidable.)

Here are examples of communication techniques which are most promising for design-decision research.

Computer graphics. Animated, computer-generated graphics are used to display the impact of construction of a new building on the view

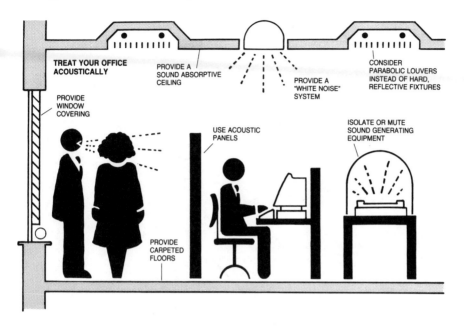

FIGURE 3. Example from a facility users' manual. From M. Brill and C. Parker, *Using Office Design to Increase Productivity for the Small Business*, Buffalo, NY: BOSTI, 1987, p. 20.

as part of a visual resource assessment. Such approaches are increasingly being used with computer systems that "capture" images of settings. These images can then be altered in the computer to reflect optional changes, and the changes can be tested for user response.

Videotape. Since videotape has become more accessible, some design-decision researchers are using it as an effective means of communications. By editing and superimposing narration, the researcher can include images from the field together with edited comments of user groups. This can be an extremely effective way of presenting responses to existing or proposed building designs. Another application was developed by BOSTI, which created a training film (Dixon, 1988) and a users manual (Brill & Parker, 1987; see Figure 3) to assist office workers in understanding the ranges of possibilities and some of the required new behaviors in a new building they were about to occupy.

Annotated schematic design drawings. Zeisel (1981) documented a presentation technique that has gained considerable acceptance with design-decision researchers—that is, to annotate the behavioral effects or implications directly on the design drawings (generally at the sche-

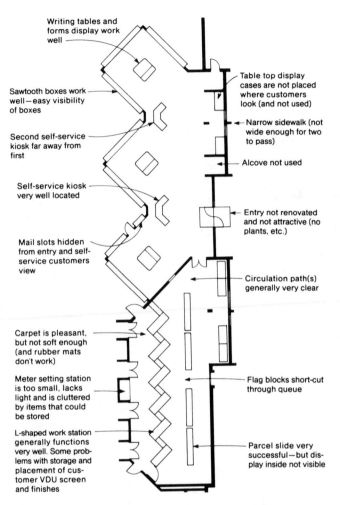

Writing tables and forms display work well

Sawtooth boxes work well—easy visibility of boxes

Second self-service kiosk far away from first

Self-service kiosk very well located

Mail slots hidden from entry and self-service customers view

Carpet is pleasant, but not soft enough (and rubber mats don't work)

Meter setting station is too small, lacks light and is cluttered by items that could be stored

L-shaped work station generally functions very well. Some problems with storage and placement of customer VDU screen and finishes

Table top display cases are not placed where customers look (and not used)

Narrow sidewalk (not wide enough for two to pass)

Alcove not used

Entry not renovated and not attractive (no plants, etc.)

Circulation path(s) generally very clear

Flag blocks short-cut through queue

Parcel slide very successful—but display inside not visible

FIGURE 4. Example of annotated plan. From E. M. Moreno, "The Many Uses of Postoccupancy Evaluation," *Architecture*, 1989, April, p. 121. Used by permission.

matic stage; see Figure 4). This is a way to communicate to some extent in the architect's own language, and to spatially locate effects for other audiences. It is also an effective way to present findings about issues and problems of a facility in use (Moreno, 1989).

Policy. In Ventre's (1988) review of the policy environment for environmental design, he mentions at least seven definable areas in which design research influences policy decisions. These are direct governmental policies, legislative mandates, government investment, regulatory policy, tax policy, judicial doctrine, and voluntary standards. Researchers are having great impact on design decisions by influencing policy. Ventre gives an example relating to standards by citing Edward Steinfeld's (1976) work on a technical basis for accessibility standards:

> The U.S. Department of Housing and Urban Development . . . supported the effort whose results were adopted voluntarily by thousands of design firms and the building owners that employed them. By the middle of the 1980s the American National Standards Institute (ANSI) reported that 95 percent of local and state building regulatory jurisdictions had adopted and were enforcing ANSI-A117.1's technical provisions, the provisions also having been adopted by reference in the three major building codes and integrated into the Uniform Federal Accessibility Standard (UFAS) of the U.S. Government. (Ventre, 1989, pp. 332–333)

WHY ARE THESE CHANGES IN DESIGN RESEARCH TAKING PLACE?

We wish to discuss two aspects of why these changes in design research are taking place. The first concerns situational factors that affect the practice of design and research, while the second concerns the diffusion of any set of ideas.

In terms of the context of design research, it is worth considering the changes described by Robert Gutman (1988) in his critical review of architectural practice (some of which were discussed above). Gutman identifies a number of trends in society and in the economy to which architectural practice has responded. Many of these would appear to cast light on changes in the practice of design research as well. Among them is the increasing demand for design services. Tied to increasing demand is the rise of the "in-house architect" within government agencies and corporations. With this in-house expertise, Gutman points out, client representatives are more sophisticated consumers of design services than typical bureaucrats or entrepreneurs, and they may well understand and be sympathetic to the needs for research-based information or consensus decision making. Gutman also discusses the evolution of larger, more sophisticated building types and an awareness of the complexity of issues surrounding construction, including economic, environmental, worker health and satisfaction, and so forth. All of these factors contribute to the demand for design research practiced as we are describing it.

The acceptance of design-decision research can also be viewed as the spread of a new idea. Diffusion of innovation theory suggests that new ideas go through stages of acceptance, passing from generation of the idea, to its development, implementation, and finally diffusion. The rate of diffusion, or adoption of the new idea, depends on six attributes (Utterback, 1974). These are discussed below in terms of their application to design-decision research.

1. *Relative advantage.* Diffusion is enhanced if there is clear relative advantage for the new idea in financial, social, or other terms. Design-decision research has become more widely adopted as decision makers begin to understand how design-decision research could help inform and/or facilitate their facility decisions.

2. *Communicability.* Diffusion is enhanced when the innovation can be explained easily and separated or identified easily. As design-decision research reports have evolved into more accessible forms, communicability, and thus diffusion, have increased.

3. *Compatibility.* If the innovation is congruent with current norms, values, or structures, it is likely to be accepted more readily. As design-decision research practitioners have become more interested in organizational structure and decision making, the congruence between client and researcher has increased, accelerating the rate of diffusion.

4. *Nonpervasiveness.* The greater the number of aspects of the organization or society that are potentially influenced by a change, the less likely it is that the change will take place. Since it is essentially pervasive, this is the one feature of design-decision research that may inhibit its diffusion.

5. *Reversibility.* Any innovation diffuses more quickly if it can be experimented with at relatively low cost of time, money, and commitment, making it easy to "back out" of the decision. Policy decisions made on the basis of design-decision research or commitments to conduct pilot design-decision research activities fit this criterion for accelerating diffusion. Actual construction does not, as it represents an expensive and long-lived investment.

6. *Small number of gatekeepers.* The fewer people involved with "keeping the gates," the greater the chance of having the innovation adopted. This trend is occurring for design-decision research. For example, some organizations with centralized design decision making, but large volume of building, have adopted design-decision research programs (such as the U.S. Postal Service, Public Works Canada, and the Marriott Corporation). In

some organizations, it has taken relatively few people, at high levels of managerial decision making, to become interested in the potential for design-decision research to help them solve their problems. Given their position and influence, they have aided the diffusion of design-decision research.

Interestingly, diffusion of innovations is usually discussed in terms of one target group of potential adopters of the new idea (for example, farmers considering a new form of plow) and in terms of an innovation being relevant to one particular aspect of a process (plowing, but not reaping). This may be thought of as "vertical" adoption, usually indicated as a rise on the vertical axis of a graph, where the horizontal axis is time. For design-decision research, the pattern has been different. Rather than being widely adopted by the original target group (architects), the approach has been picked up by others (facility managers, business leaders, policymakers, regulatory bodies, and organizational planners). In a similar departure from diffusion of innovation theory, the original concept of design-decision research influence has been spread across a wide range of aspects of the process.

IMPLICATIONS FOR FUTURE RESEARCH AND APPLICATIONS: WILL WE STAY IN THE SWAMP?

Swamps are uncomfortable places, sticky and full of unknown or unrecognized creatures. We fear them because of their ambiguity and unfamiliarity. While unfamiliarity may recede over time, ambiguity does not. We would suggest that ambiguity, if uncomfortable, is not all bad; it is in fact a motivator for learning. In discussing "reflective thinking," John Dewey (1933) says that it involves "(1) a state of doubt, hesitation, perplexity, mental difficulty, in which thinking originates, and (2) an act of searching, hunting, inquiring, to find material that will resolve the doubt, settle and dispose of the perplexity" (p. 12). This is a way of thinking about creativity; about the creative aspects of design-decision research and research-based design. Rather than seeing the swamp as dark and frightening, we find it full of richness and possibility.

As we have made clear above, we perceive design-decision research to be a mode of "application" with great future potential. However, there are several ways in which we can learn to improve the practice of design-decision research and perhaps aid its diffusion. One is to document case studies of successful design-decision interventions. While we have cited some above, few are documented in a way that allows the field to learn about the process. Such case studies would also be valuable in formal design education.[1]

Going beyond simple documentation would be in-depth studies of one (or a family of) design-decision research examples. These studies should begin to draw out the patterns within the process and establish factors and preconditions contributing to degrees of success or failure. An example of this is Carpman's (1983) study of participation in a hospital planning project. We need to understand much more about the swamp and its creatures—so that design-decision research can become more effective and efficient.

REFERENCES

American Institute of Architects (1980). Owner-Architect Agreement. *Handbook of Professional Practice*. Washington, DC: Author.

Archea, J., & Margulis, S. J. (1979). Environmental research inputs to policy and design programs: The case of preparation for involuntary relocation of the institutionalized aged. In T. O. Byerts, S. C. Howell, & L. A. Pastalan (Eds.), *Environmental context of aging: Life-styles, environmental quality, and living arrangements* (pp. 217–228). New York: Garland Press.

Bechtel, R. (1978). *Post occupancy evaluation of housing*. Washington, DC: U.S. Department of Housing and Urban Development.

Brill, M., & Parker, C. (1987). *Using office design to increase productivity for the small business*. Buffalo, NY: Buffalo Organization for Social and Technological Innovation.

Carpman, J. (1983). *Influencing design decisions: An analysis of the impact of the patient and visitor participation project on the University of Michigan replacement hospital project*. Unpublished doctoral dissertation, University of Michigan, Ann Arbor.

Carpman, J., Grant, M., & Simmons, D. (1986). *Design that cares: Planning health facilities for patients and visitors*. Chicago: American Hospital Publishing.

Carpman, J., Grant, M., & Simmons, D. (1984). *No more mazes: Research about design for wayfinding in hospitals*. Ann Arbor: University of Michigan, Office of the Replacement Hospital Program.

[1]We offer the following comments on design education, which derive from our theses about design research. We believe that design education needs to go further in recognizing information-based design decision making as representative of the real world. This is beginning through increasing use of case study methods (Joroff & Moore, 1984; Symes, 1989). The recent efforts of ACSA to sponsor the development of case study materials as the bases for design studio curricula is another step in the right direction. While these case studies (with the exception of some of Symes's work) do not directly refer to environment–behavior research as information input, any case method which reflects how decision-makers are influenced by information is a good beginning. Developing cases which have environment–behavior research-based information as a part of the decision process will be a useful pedagogical tool.

Dewey, J. (1933). *How we think.* (rev. ed.). New York: Heath.

Dixon, J. M. (1988). Augmenting design through user training. *Progressive Architecture, 1* (January), 143.

Farbstein, J., Kantrowitz, M., Schermer, B., & Hughes-Caley, J. (in press). Post-occupancy evaluation and organizational development: The experience of the United States Postal Service. In W. F. E. Preiser (Ed.), *Building evaluation: Advances in methods and applications.* New York: Plenum.

Farbstein, J., Archea, J., Kantrowitz, M., Shibley, R., Wineman, J., & Zimring, C. (1986). Designing and building with rehabilitation in mind. In G. Davis (Ed.), *Building performance: Function, preservation and rehabilitation* (Special Technical Publication 901, pp. 39–45). Philadelphia: American Society for Testing and Materials.

Gutman, R. (1988). *Architectural practice: A critical review.* Princeton: Princeton Architectural Press.

Hart, G. K., & Kurtz, J. (1980). Designers are the first market for passive solar innovation. *Solar Review.* Washington, DC: United States Department of Energy.

Jockusch, P. (in press). Post-occupancy evaluation as a tool for the preparation of architectural competitions. In W. F. E. Preiser (Ed.), *Building evaluation: Advances in methods and applications.* New York: Plenum.

Johnson, D. W., & Johnson, F. P. (1982). *Joining together: Group theory and group skills.* Englewood Cliffs, NJ: Prentice Hall.

Joroff, M., & Moore, J. (1984). Case method teaching about design process management. *Journal of Architectural Education, 38*(1, Fall), 14–17.

Kantrowitz, M. (1985). Has environment and behavior research "made a difference?" *Environment and Behavior, 17,* 25–46.

Koncelik, J. A. (1982). *Aging and the product environment.* New York: Van Nostrand Reinhold.

Min, B.-H. (1988). *Research utilization in environment-behavior studies: A case study analysis of the interaction of utilization models, context, and success.* (Doctoral dissertation, Department of Architecture, University of Wisconsin-Milwaukee). Ann Arbor, MI: University Microfilms.

Moleski, W., & Lang, J. (1982). Organizational goals and human needs in office planning. In J. Wineman (Ed.), *Behavioral issues in office design* (pp. 3–21). New York: Van Nostrand Reinhold.

Moreno, E. M. (1989). The many uses of postoccupancy evaluation. *Architecture, 78*(4), 119–121.

National Endowment for the Arts (1984). The geriatric personal furnishings system. *Exemplary design research, 1983.* Washington, DC: Author.

Pauls, J. (1986). Mock trial: Litigation following an injury in a fall on a residential stair. In J. Wineman, R. Barnes, & C. Zimring, (Eds.), *The costs of not knowing* (p. 387). Washington, DC: Environmental Design Research Association.

Preiser, W., Rabinowitz, H., & White, E. (1988). *Post-occupancy evaluation.* New York: Van Nostrand Reinhold.

Schneekloth, L. H. (1987). Advances in practice in environment, behavior, and design. In E. H. Zube & G. T. Moore (Eds.), *Advances in environment, behavior, and design* (Vol. 1, pp. 307–334). New York: Plenum.

Schon, D. A. (1984, April) Education for reflection-in-action. Paper presented at the Harvard Business School Anniversary Colloquium on Teaching by the Case Method, Cambridge, MA.

Steinfeld, E. (1986). A case study in the development of a research-based building accessibility standard. In J. Wineman, R. Barnes, & C. Zimring (Eds.), *Proceedings of the seventeenth annual conference of the Environmental Design Research Association*, Washington, DC: EDRA.

Stivers, E., & Wheelan, S. (Eds.). (1986). *The Lewin legacy: Field theory in current practice*. Berlin: Springer Verlag.

Symes, M. (1989). The case method: A paradigm for environmental design research. Unpublished manuscript, University College London, School of Environmental Studies.

United States Postal Service. (1987). *Retail design guidelines*. Washington, DC: Author.

Utterback, J. M. (1974). Innovation in industry and the diffusion of technology. *Science, 183,* 620–626.

Ventre, F. T. (1988). The policy environment for environment and behavior research. In E. H. Zube & G. T. Moore (Eds.), *Advances in environment, behavior, and design* (Vol. 2, pp. 317–342). New York: Plenum.

Vischer, J. (1989). *Environmental quality in offices*. New York: Van Nostrand Reinhold.

Wandersman, A. (1980). Combining research and practice in citizen participation. In R. R. Strough & A. Wandersman (Eds.), *Optimizing environments: Research, practice, and policy* (p. 103). Washington, DC: Environmental Design Research Association.

Woods, J. H., & Casper, I. G. (1986). Using graduate students as consultants to teach action research to residence hall staff. In E. Stivers & S. Wheelan (Eds.), *The Lewin legacy: Field theory in current practice* (pp. 158–165). Berlin: Springer Verlag.

Zeisel, J. (1975). *Sociology and architectural design*. New York: Russell Sage Foundation.

Zeisel, J. (1981). *Inquiry by design: Tools for environment-behavior research*. Monterey, CA: Brooks/Cole.

Zeisel, J. (1986). Building purpose: The key to measuring building effectiveness. In M. Dolden & R. Ward (Eds.), *The impact of the work environment on productivity: Proceedings of a workshop*. Washington, DC: National Science Foundation and Architectural Research Centers Consortium.

Zimring, C., Wineman, J., & Carpman, J. (1988). The new demand-driven post-occupancy evaluation. *Journal of Architectural and Planning Research, 5*(4), 273–283.

10

"Einstein's Theory" of Environment– Behavior Research

A COMMENTARY ON RESEARCH UTILIZATION

DAVID KERNOHAN

THE "THEORY"

Einstein's theory of relativity convinced us of the importance of the fourth dimension—time. In turn, explorations in quantum theory led to Heisenberg's uncertainty principle, which is "a fundamental, inescapable property of the world" (Hawking, 1988, p. 55). Heisenberg's principle showed that the more accurately attempts to measure the position of a particle are made, the less accurately its speed can be measured, and vice versa. Therefore, there cannot be a theory of science, a model of the universe, that is completely deterministic. One cannot predict future events if one cannot even measure the present stage of the universe precisely.

Einstein and Heisenberg presented a model of the world that is dynamic and uncertain. Environmental designers and researchers operate in such a world—if not at the level of quantum physics, certainly in the day-to-day interactions of people and the physical environment. They operate in environments where change and uncertainty prevail.

David Kernohan • School of Architecture, Victoria University of Wellington, Wellington, New Zealand.

Yet the manner of their operation usually belies the dynamics of the situation. Traditionally environmental design and research have operated from a Newtonian model that is static and deterministic. Schon's "hard, dry ground . . . of well formed problems" is the territory of this traditional approach (see Farbstein & Kantrowitz, Chapter 9, this volume, or Schneekloth [1987]). The traditional approach is characterized by the separation of researcher and designer. It is where the "two communities" (Min, 1988) reside. The swampy terrain below, where, according to Farbstein and Kantrowitz, "pressure for answers is constant and professional boundaries are often unclear," may be the habitat of the "one-community" person, the practitioner in both design and research.

THE "GAP"

In this book, the chapters of Wisner, Stea, and Kruks (Chapter 8) and of Farbstein and Kantrowitz (Chapter 9) reflect advances and changes in our understanding of the relationships of environment–behavior research and design and of environmental design and research. Much of what is discussed is concerned with the purposes of research, who it serves, how it is conducted and applied. Both chapters are concerned with applied design research practice, that is, the application of research methods, processes, findings, and outcomes to directly inform decision making in contexts of environmental change. They record and discuss advances and changes which generically stem from action and participation research. Their foundation is a belief that "to be effective, people involved in social change must also be involved in the process of generating knowledge about that change, in posing issues to be researched, in implementation, and in evaluation" (Lewin, 1946, p. 34).

Wisner, Stea, and Kruks explore the role of the researcher in "research for the people/research by the people." They compare Lewin's view of action research with the intellectual movements of the 1920s, 1930s, and 1940s in Latin America and Asia and link both to the academic fields of adult and continuing education and the dialogue between university-based researcher and rural people. They emphasize that the "actor" in action research is the researcher. "In participation, emphasis is placed on the participants as primary actors." Farbstein and Kantrowitz introduce the term "design-decision research." They suggest that while action research was "a radical departure from the 'isolated researcher' prevalent prior to Lewin's work, it still assumes that research is not integral to change, but is separate from it in time and person."

Farbstein and Kantrowitz argue that design research within the traditional applied research model is divorced in time and persona from the arena of design decision making. Like design, which is traditionally modeled as a production system with a design process, clients, users, completed buildings, and their operation and maintenance as separate entities, design research is essentially compartmentalized and static. They cite the consultative model, where researchers perform an analysis and present results separate from the designer who may use them. They acknowledge that design research "still is far from fully integrated into standard design practice." This is an understatement, for until a new paradigm takes hold, it is here that the "gap" and the "squishy middle" between research and design remains.

The major problem which faces environment-behavior research is how to make its findings more usable in other fields such as environmental design and management. The methods and outcomes of environment–behavior research are well documented. It is a field of study that offers much academically and intellectually yet its methods, processes, findings, and outcomes are only rarely applied in design practice. Environment–behavior research and design and management practice are two different and currently disparate activities. There are three principal reasons for this.

The first is that much of the activity in environment–behavior research, as its title indicates, is research oriented. It is concerned with researching environment–behavior relationships in the classic scientific mode with a view to discovering laws about them and forming theories. This may be a limiting approach.

> The origins of this widely prevalent model of behavioural research lie in the application to the study of human social behavior of the same rules which the Newtonian model of science applies to the study of physics. . . . There is a binding assumption that somewhere out there, there are fixed and immutable laws linking environment and behaviour to be found. The possibility that the quest may be illusory and that there are actually rather few generalities of any significance about environmental behaviour to be found has not taken hold. (Ellis, 1983, p. 430)

The second reason relates to the nature of academic life itself. For all the amalgam of disciplines in EDRA, IAPS, MERA, and PAPER, the cooperation and collaboration implied by these umbrella organizations rarely repeats itself in either the research or practice domains. It is not two communities of researchers and designers but many communities, discrete and compartmentalized. The professional distinction of psychologist and sociologist is as strong as that of architect and engineer. While there are examples of researchers and practitioners working effectively together, they are few. More generally, they do not collaborate. Each operates from a different mind-set.

The third reason has to do with that mind-set. There are significant differences between the aims and activities of researchers and those of designers. Research is concerned with the generation of knowledge. Design needs knowledge to inform its pragmatic decision making. However, research knowledge is rarely shaped or sorted to a usable form for specific practice contexts. It resides uninterpreted in academic reports that designers and managers tend not to read, to refer to, or in most cases even to know about.

> Information transfer is based on the understanding that knowledge is generated in one sphere, sent out into the world, and picked up and used by others who are potential consumers of research results. This form assumes a noninterventionist stance. It uses scientific conventions on the one hand, and the application of knowledge on the other. The relationship between the generation of knowledge and its subsequent use is discontinuous and the time frame indeterminant. (Schneekloth, 1987, p. 309)

Exacerbating the difficulty is the way in which the design professions are assailed on all sides by information: new construction methods, materials, technologies, techniques; new regulations, codes, standards, guidelines, research. Information technology does not address how the individual is to assimilate or use all the available information.

Building design is not a literary activity. A study of design decision making in architectural practice showed that:

> Designers based their decisions largely on personal and practice experience and that they used few publications. . . . Any information that designers consult must be quick and easy to absorb. . . . Manufacturer's were the main source consulted; they were preferred to official publications. . . . The research gave a strong indication that designers seek written information as a last resort. (Mackinder & Marvin, 1982, p. 8)

Personal experience, aesthetic doctrine, and normative data enable designers to proceed in a world of information overload and uncertainty. Environmental designers and managers seek the parameters within which they may operate. Rules, regulations, and codes are ideal. They spell out what can and cannot be done. Bits of prescriptive advice (rules of thumb) help. Building performance research sometimes lends itself to providing rules of thumb, but behavioral research rarely can do so. Relying on personal experiences, doing what one is accustomed to or what has caught the eye, and arbitrary and cursory reference to pertinent research are, without enhancement, grossly inadequate tools to face a rapidly changing future. They are all manifestations of the static model of current design and management practice. There is good argument that a new paradigm integrating research and design is required to deal with the phenomena of change in the relationships of people, organizations, and their buildings.

A NEW PARADIGM

According to Farbstein and Kantrowitz, design-decision research is part of a new paradigm that addresses the context of design. It is part of a continuum that allows design and research to be performed simultaneously. It is a different way of thinking and acting in relation to the design process. "It focuses explicitly on helping clients realize their objectives. . . . The researcher's role is to help the organization make its own best decisions within the context of its objectives." Its outcomes affect decision making directly. Farbstein and Kantrowitz recognize that who the "client" is affects how such research can be carried out.

Wisner, Stea, and Kruks approach the same issue through their discussions of participation and participatory research. They discuss participation as action, or praxis, which stresses the process of participation itself. They also discuss participation as communication and as research and review techniques of participation. They accept that participation may not harmonize but can in fact foster conflict. They draw a distinction between and provide examples of "instrumental" and "transformative" participation. The former is concerned with efficiency and effectiveness, while the latter "is seen to facilitate changing social consciousness."

Our own research (Kernohan, Daish, & Gray, 1987) has found that participatory evaluation is a key to the successful negotiation of environmental design quality. This is based on the belief that environmental design cannot be modeled as a production system but

> can be described more adequately, if less clearly, as a social process. . . .
> Social experience is continuously accumulating and changing, and so meaning in environment is negotiable and temporary. Design and operational quality is therefore negotiable." (Ellis & Joiner, 1985, p. 130)

This has led to the development of a generic evaluation method that is used for a range of predesign, design, and postoccupancy applications. The method enables various interest groups to record and discuss their perceptions of issues in a proposed or existing environment. It then provides a negotiation process that leads to recommendations for action to affect that environment.

Our procedures operate in the swampy terrain. Our roles as researchers are certainly muddied. The evaluations are carried out by those with interests in the building or building project. A neutral task group facilitates the evaluations. Task group members may be researchers but are more often design and management practitioners prepared for the evaluation event through guidance documentation or in consultation with researchers. In this research much effort is expended toward empowering others to undertake their own participatory research. As Farbstein and Kantrowitz suggest, research in this context is as concerned with process as with generating information for design decision making.

The outcomes of such evaluations are not reports "to sit on someone's shelf." They are actions in response to the recommendations of the participants in the evaluation. Sometimes these actions include specialist evaluations by expert researchers. Are such evaluation activities research, design-decision research, or are they enhancements of the normal analysis processes of designers?

For such organizational interventions to succeed they must be sympathetic to the host culture. Participatory evaluations, while contributing to the organizational processes of democratic organizations, may be alien to others. Wisner, Stea, and Kruks stress that participation itself need not necessarily imply a democratic process. While participation can be a manifestation of the ongoing struggle by people to control their own lives, it can also be initiated and utilized by external power holders. The issue of communication is raised: what is communicated, who does the communicating, and why? They indicate that "noise" influences how messages are transmitted and received and refer to Foucault's belief that the inseparability of power and knowledge precludes all possibility of neutral relations between research/designers and client/participants. They question the possibility of co-equal participation of citizen and researcher alike.

THE CAUTIONARY TAIL

Farbstein and Kantrowitz suggest design-decision research is taking hold because of an increasing demand for design services—a demand which is coming not from architects, the "original target group" of design-decision research, but from "others (facility managers, business leaders, policymakers, regulatory bodies, and organizational planners)." They indicate that the practitioners of design-decision research are not an identifiable group. Some are from academic backgrounds but many operate under job titles such as those listed.

Certainly facility managers as one group are in the market for design-decision research. Currently they lack tools to directly address issues of user satisfaction, functionality, and productivity, in order to negotiate ongoing design quality. While they are well served in areas of asset and inventory management and cost accounting through computer-aided facilities management, they are less well served with methods for addressing the less readily quantifiable needs of an organization's most valuable asset—its people. Facility managers are familiar with the phenomenon of "churn," the proportion of the workforce that changes every year, but they are unsure how to address some of the issues it raises. Design-decision research and participatory evaluations can assist, not just as management techniques, but by redefining the role of research in relation to environmental design and management activities.

Less positively, but nevertheless significantly, reductions in research funding in a number of countries have led researchers themselves to research which, while "consciously directed toward contributing to design decisions," may also be directed to the specific requirements of the paying client. This reopens Wisner, Stea, and Kruks's concerns about the politics of participation and participation research. There are many unseen dangers in the swamp. While it may feel like the right place to be, some caution is necessary.

> Any discipline needs a range of activities from problem solving in the front lines by practitioners, through efforts by methodologists and theoreticians to detached contemplations by philosophers." (Rapoport, 1979, p.15)

Bechtel (1987) has rightly warned that some methods in the swamp can be too "quick and dirty." If progress is to be made, then intellectual rigor must be maintained. The chapters by Farbstein and Kantrowitz and by Wisner, Stea, and Kruks indicate that those who are entering the swamp are aware of the issues as they step. Both are concerned with who is being served and who benefits from design research interventions. They question "the relationship between managerial 'sponsorship' of research and those with the power to implement findings and recommendations," and the role of "instrumental" and "transformative" participation.

Whether on the "hard, dry ground" or "in the swampy terrain," whether operating in a traditional academic research role or in the ongoing negotiation of design quality, we operate in a world of uncertainty and change. There is now a better understanding of environmental design and management as a social process that involves many people in the ongoing negotiation of design quality. This will continue to lead to greater integration of research and designing activities. Nevertheless, it will be important to retain some people on the hard dry ground—if only to ensure that no one is sucked under.

REFERENCES

Bechtel, R. (1987). Advances in POE methods. In N. Praks (Eds.), *Proceedings of the International Association for the Study of People and their Physical Surroundings* (Vol. 2, p. 350). Delft, the Netherlands: Technical University of Delft.

Ellis, P. (1983). *Institutional problems with design research in Britain.* In D. Joiner, G. Brimilcombe, J. Daish, J. Gray, & D. Kernohan (Eds.), *Proceedings of the conference on People and Physical Environment Research* (PAPER; p. 566). Wellington, New Zealand: Ministry of Works and Development.

Ellis, P., & Joiner, D. A. (1985). *Design quality is negotiable.* In K. Dovey, P. Downtown, & G. Missingham (Eds.), *Place and placemaking: Proceedings of the conference on People and Physical Environment Research* (PAPER). Melbourne, Australia: Melbourne University.

Hawking, S. W. (1988): *A brief history of time.* New York: Bantam Books.

Kernohan, D., Daish, J. R., & Gray, J. A. (1987). *Building evaluation: Data gathering and review.* In B. Shaw (Ed.), Conference on People and Physical Environment Research, Perth, Australia: Centre for Urban Research, University of Western Australia.

Lewin, K. (1946). Action research and minority problems. *Journal of Social Issues, 1-2,* 34–36.

MacKinder, M., & Marvin, H. (1982) *Design decision making in architectural practice.* Garston, England: Building Research Station, BRE Information Paper IP 11/82.

Min, B. H. (1988). *Research utilization in environment-behavior studies: A case study analysis of the iteration of utilization models, context, and success.* (Doctoral dissertation, Department of Architecture, University of Wisconsin-Milwaukee.) Ann Arbor, MI: University Microfilms International.

Rapoport, A. (1979). Pure research in architecture and urban planning. *Wisconsin Architect, Journal of the Wisconsin Society of Architects.* Madison, Wisconsin.

Schneekloth, L. H. (1987). Advances in practice in environment, behavior, and design. In E. H. Zube & G. T. Moore (Eds.), *Advances in environment, behavior, and design* (Vol. 1, pp. 307–334). New York: Plenum Books.

Contents of Previous Volumes

Index